THE PULPWOOD QUEENS' TIARA-WEARING, BOOK-SHARING GUIDE TO LIFE

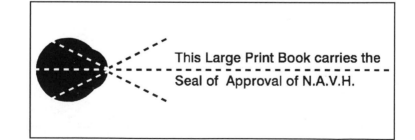

This Large Print Book carries the
Seal of Approval of N.A.V.H.

THE PULPWOOD QUEENS' TIARA-WEARING, BOOK-SHARING GUIDE TO LIFE

KATHY L. PATRICK

THORNDIKE PRESS

A part of Gale, Cengage Learning

Detroit • New York • San Francisco • New Haven, Conn • Waterville, Maine • London

Copyright © 2008 by Kathy L. Patrick.
Prologue, "I Want My Hair Considered," written by Marshall Chapman and Kathi Kamen Goldmark is copyright © 2005 Tall Girl Music (BMI). Administered by Bug/Don't Quit Your Day Job Music (BMI). Testimonial on page 222 copyright © 2007 Robert Dalby; and on pages 262 and 263 © 2005 Marshall Chapman.
Thorndike Press, a part of Gale, Cengage Learning.

LIBRARY OF CONGRESS CATALOGING-IN-PUBLICATION DATA

Patrick, Kathy L.
  The Pulpwood Queens' tiara-wearing, book-sharing guide to
life / by Kathy L. Patrick. — Large print ed.
    p. cm.
  ISBN-13: 978-1-4104-0782-5 (hardcover : alk. paper)
  ISBN-10: 1-4104-0782-9 (hardcover : alk. paper)
  1. Pulpwood Queens Book Club. 2. Book clubs (Discussion
groups) — Texas. 3. Books and reading — Social aspects —
Texas. 4. Patrick, Kathy L. I. Title.
LC6652.T48P38 2008
367.9'764—dc22   2008011079

Published in 2008 by arrangement with Grand Central Publishing, a division of Hachette Book Group USA, Inc.

Printed in the United States of America
1 2 3 4 5 6 7 12 11 10 09 08

In memory of
Jonathan Douglas-Sanders

ACKNOWLEDGMENTS

If someone had told me, a small-town girl from Kansas, that someday I would write and publish a book on my life, I would have told them, "Get out of here!" One true thing I do know is that if it had not been for the following people who have encouraged me and helped me along my life's journey, none of my accomplishments would have been possible.

First, I must thank my dear, dear friend and agent Marly Rusoff. I will never forget our first meeting by phone. She asked, "What is your favorite book, Kathy?" I told her it is *To Kill a Mockingbird* by Harper Lee. "Me too!" was her reply, and that began our lifelong relationship of sharing our love of author Harper Lee, Southern fiction, and good books. To have someone believe in you as much as Marly has believed in me is something that I will treasure forever. She has invested in me monetarily and in mind, body, and spirit! She has held my hand on many occasions and given me a shoulder to cry on. Some people you can never repay, but I will, to my dying day, be thankful for knowing such a caring, loving, savvy book lover as Marly. And, Marly, I must substitute agent for angel. You have to know that you are solely responsible for bringing this book to life! X's and O's!

Next, I have to thank Bob Summers. He is the one who suggested — no, insisted — I write my story and continues to lift me up to places I never dreamed. My life has been blessed for having such a fellow book enthusiast and mentor lift me up. What a friend you are indeed.

For Amy Einhorn, who loved my story on Pulpwood Queen Joyce Jackson Futch enough to purchase my book for Grand Central Publishing, my gratitude and thanks.

To my editor, Natalie Kaire: You have been a dream of an editor. You have made my book one that I am proud of writing and I thank you for all your hard work to polish this ol' rough lump of coal into a diamond. You sure had your work cut out for you.

To Elly Weisenberg, I thank you for giving me the insight and proper direction to make my diamond shine. You are a true gem!

8

In my book, I have mentioned many people who believed in me when I did not even believe in myself. Here are more unsung heroes:

To the late Andrew and Warda Hibbard and family: You all took me into your fold and opened my eyes to the power of education, literacy, politics, lively conversation, heated debate, and belief in the human spirit. My days and nights with your family of lawyers, journalists, and educators are what opened my world to the possibility that I could do anything if I educated myself and worked hard enough.

To my former college professor Gary Marsh: Your taking me under your wing, your encouragement, and your kindness will always be remembered. You must know that you and a simple note you once wrote me have kept me going when sometimes life's journey seemed more than I could bear.

To my living and breathing Tarzan, aka Geoff Cresswell, my college friend, whose sardonic wit, intellect, and perfectionism have continued to make me laugh through the years when I felt like crying and throwing in the towel. Um ga wa from your Conan Barbie!

I also have to thank my first book family, Jim Barron and the wonderful staff of Barron's Books! You were my first informal "book club" and I never could have accomplished what I have without your giving me that first chance at bookselling and educating me on the wonderful world of books. Our time together will always be cherished and held close to my heart!

To bookseller extraordinaire and my mentor, Mary Gay Shipley: Your humor, guidance, and friendship are valued deeply. You are the Kat's *meow!*

To the late Jim Shepherd, who truly took a chance and gave me my first dream job. Who knew our book world would get knocked off its axis?

To John Berendt, who gave me wings and let me fly. Your encouragement was what it took, for me, to leave my safe little nest.

To Pat and Cassandra Conroy: Pat gave me my best author experience I have ever had and my first experience as a journalist. To meet my hero in person was a dream come true. Cassandra, you complete the picture. You are my favorite authors and couple, and your support and encouragement have been incredible.

To Christopher Cook, who has been a great friend and mentor. He is my "book whisperer" and the one person that I can turn to for anything.

To Diane Sawyer and Charlie Gibson, who showcased our book club when we helped kick off their Read This Book Club on *Good Morning America.* I probably would still be running my one lone book club in Texas if it had not been for them. Diane, you truly crowned me Queen, and Charlie, you stood up for this southern hairdresser/bookseller when I was questioned on what could anyone with *big hair* really know. Evidently, thanks to you both, a lot.

I have been featured in many newspapers, magazines, and television shows but if you want to know what meant the most to me personally? That my former Sunday school students, whom I taught from third grade on up to teenagers, now adults, all got together to watch me on *Good Morning America.* That made me cry in a good way. I love you all forever. Now, who can still go roller skating to celebrate with this fifty-year-old woman?

When I started the Pulpwood Queens of East Texas, I never dreamed it would grow into the largest "meeting and discussing" book club in the world. I just wanted to be in a book club. To my darling Pulpwood Queens: Each one of you is special to me, and my life would not be complete without the friendship of each and every one of you.

11

This is not just my story; it is your story too, unfolding as I write.

To my Pulpwood Queen "Splinter" and "Pinecone" chapters: You kids have worked your tails off volunteering to help me with my author events and book festivals. Hopefully, this book will sell so many copies that I can give you great scholarships to your continuing education. Pay it forward, kids, pay it forward.

To all the authors who have graced my doors and the pages of this book: Your books were what saved me, and for that, I will be forever grateful. I hope your words and stories live forever.

To James Patterson, who generously started the James Patterson PageTurner Awards and made me one of the recipients of that award to help all of us literacy crusaders on our mission. He actually took time out of his incredibly busy schedule to read and blurb my book, what a gift!

To my generous corporate sponsors, including International Paper, Redken, Armstrong McCall, Farouk Systems, Bed-Head, Woody's products, the *Marshall News Messenger,* and Cox Newspapers: Thank you for supporting my literacy endeavors.

To all my wonderful teachers at Eureka ISD and the librarians of Eureka Carnegie

Library: You may never know what a difference you have made in my life. Since then I have had many outstanding educators as I continue my lifelong learning. You may never be thanked personally, but I want you to know, you changed my life for the better.

To all booksellers and book reps everywhere: I know that we have as much choice in selling books as those who are called to preach God's word. We are called to serve and spread the love of reading. All we ask in return is that we get to read the books.

To my fellow hairstylists, Marla Keith and Nelson Collier, you two have to put up with my eccentric ways and total lack of organizational skills on a daily basis. You make me look good, both inside and out! And to all salon professionals: We spend so much time making other people look good, let's help the world know that we all do a great service to our communities; we have value, as we serve others. There is no higher honor.

To Fred McKenzie, my fellow bookseller: You are probably the first person I have ever met who shares exactly the same beliefs in politics, religion, and life in general as I do. You may be eighty-nine-years-old, but you are my confidant and best friend. I love you dearly.

To Phil Latham, thank you for seeing the

big picture and helping me get the word out that reading is important. You are a literacy leader and a great friend.

To the Jefferson Rotary Club for welcoming probably the craziest business women into their club. I love all of you to pieces, service above self.

In order for me to know my purposeful life, I thank my God above and my spiritual leaders, Reverend Leigh, Reverend Daniel, Pastor Jerry, Preachergirl Pastor Polly, and all your families. We must put God first in all we do, and you were all there to guide me.

To Michael and Melanie Morris, Charles and Christy Martin, Denise Hildreth and Jonathan Pierce for being my Christian book family! Your love and joining my book-loving fray has made you part of my inner family circle.

Last, but most important, is my family: Mother, Daddy, Karen, Richard, Susie, Chance, and Colton. I had to tell my story. This was not a choice for me but a calling. I love you all and hope you will understand that telling my story has brought me full circle. I now understand that the only way you can truly love someone is to love yourself first. I can now say I truly love you.

And to my cousin Terry, who has always

been my shining star, my beacon of hope. I watched you strive for the highest in education and always have a strong and beloved family life; you are my hero.

To our two precious and precocious Rotary Exchange students, David Ley and Chisato Sayama! Oh, the tales you must have told when you went back home to Germany and Japan! I can't wait to read your books.

To my immediate family, my life: Jaybird, Lainie, and Madeleine, you are everything to me and what I hold most dear and close to my heart. Nothing would mean anything without you by my side. You are my past, present, and future! You must know that because of you, I have truly found my life's purpose. I love you with all my heart, body, and soul! Now maybe, Jay, you will read my book. And for those reading this, he has always left the room when I tried to read a chapter aloud. As he has told me again and again, "Why read the book, Kathy? I have lived it!"

CONTENTS

PROLOGUE

I Want My Hair Considered

Hello my name is Holly
East Texas is my home
I have a lot of girlfriends
I talk to on the phone
Tonight we're gonna dress up
And go out on the town
I need my hair and makeup done
So I won't let them down

Mildred get your hairspray
And do me up just right
I want my hair considered
At the Pulpwood Ball tonight

My shoes are pink and pointy
My dress is pink and black
I bought them at a closeout sale
So I can't take them back
My feather boa sparkles
From my neck down to my toes

But the total look I'm going for
Requires more than my clothes

So Mildred get your hairspray
And do me up just right
I want my hair considered
At the Pulpwood Ball tonight

Now step back all you beauty queens
And make some room for me
I'm gonna have the biggest hair
The world has ever seen
I may not win the prize tonight
But I don't really care
'cause everyone's a winner
When you've got outrageous hair

So Mildred get your hairspray
And do me up just right
I want my hair considered
At the Pulpwood Ball tonight
I want my hair considered
At the Pulpwood Ball tonight

— Kathi Kamen Goldmark and
Marshall Chapman

CHAPTER 1
IF LIFE HANDS YOU A LEMON, MAKE MARGARITAS

"In the beginning there was nothing. God said, 'Let there be light!' And there was light. There was still nothing, but you could see a whole lot better."
— *Ellen DeGeneres*

If you saw me today in my full Pulpwood Queens Book Club regalia featuring hot pink, leopard skin, and a diamond tiara, you might not immediately think of me as a bookseller. You might think "hairdresser" first. And you would be right on that score. But the truth is that I am both. What you have here is my story, for better or for worse.

I never knew what I wanted to be when I grew up. I ambled along, taking my lead from others. My father had run the city pool, so my sisters and I were taught to swim as babies. I took swimming lessons from kindergarten to high school, and during my high-school years I worked during the summer lifeguarding at the Eureka Country Club. Since I had never heard of anyone becoming a professional lifeguard (this was years before the television show *Baywatch*), I saw lifeguarding simply as a way of getting a great tan. Very important in the 1970s. I socked away all my paychecks toward my savings account for college (my parents had informed me from the time I was knee high to a tadpole that I was going to go).

Now, I did not have a clue as to what I was going to be in life. I just knew my mother insisted — no, commanded — that I get an education. When I graduated from high school, I did what most of my friends did: I went to college. I enrolled at Kansas State University in Manhattan, Kansas, home of the K-State Wildcats.

I attended college at K-State without much direction. After a few years, I dropped out to attend Crum's Beauty College in Manhattan, Kansas. At the time, I figured I could become a hairdresser and then go back to school when I figured out what I really wanted to be in life. I worked in several college salons in both Manhattan and Lawrence, Kansas, before I moved back to my hometown in Eureka to open my first salon, Town & Country Headquarters. Named after my favorite magazine, *Town & Country,* I thought the name was brilliant because Eureka was the county seat so I did both town folks' and country folks' hair. Unfortunately, I was always getting calls asking, "Do you sell John Deere or International Harvester tractors?" Callers were very surprised when I told them my shop was a beauty shop, though it was still a far cry from the glamour and elegance of that magazine.

Eventually I closed down Town & Country Headquarters and spent the next few years in and out of a few colleges, working in several hair and makeup jobs, and even moving out to California — forever searching for my life's work.

I found California absolutely beautiful but I missed my family, my friends, and the weather back home. Where is a good thunderstorm when you need one to curl up in front of the fireplace and read a good book? Unable to face another Christmas spent going to a double feature at the movies and having my holiday feast be a couple of hot dogs, I made up my mind. I was heading to Texas for a real Christmas to visit my sister.

> "Texas is a state of mind. Texas is an obsession. Above all, Texas is a nation in every sense of the word. And there's an opening covey of generalities. A Texan outside of Texas is a foreigner."
> — *John Steinbeck, American novelist*

My sister Karen was living in historic Jefferson, Texas, which is set on the banks of Big Cypress Bayou. Karen and her husband, Richard, had recently purchased a fixer-upper Victorian home built right after the turn of the century. I could not wait to see it, and being with them was always a very special time. Needless to say, my plan for a holiday visit turned into something weird and totally unexpected: I fell in love with Jefferson, Texas!

Two weeks later, I had moved to Texas and purchased a historic home built Louisiana-steamboat-style right next to my sister. That following May, I opened Heart of Pine Cottage Bed & Breakfast, named after the home's beautiful heart-of-pine floors.

I ended up marrying a hometown boy. After a whirlwind romance, we eloped (which could be a story in itself). Then a couple years later, after having our first

24

child, I decided I was going to pursue a job that was — besides hair — my other true passion. I love to read and I love books. I decided that I wanted to work in a bookstore.

Starting as a lowly bookseller, I began working from the ground up in a nearby independent bookstore. Several months later, I was promoted to children's manager and buyer. I read everything I could get my hands on — books, trade magazines, and book-publisher catalogs. I absorbed the book business like a sponge. I lived and breathed books and began to feel as if printer's ink was coursing through my veins. I commuted back and forth to the bookstore even through my second pregnancy, and worked there for most of my children's formative years. Then I was made an offer I could not refuse. I accepted the job of a book lover's lifetime: I became a book publisher's representative.

I loved working in the bookstore. Not only were books my friends, but that bookstore and my coworkers became my second family. I compare those years at the bookstore with my first twelve years of schooling and being raised by my family. I had learned everything I possibly could, but now, with my new job offer, it was time to graduate.

Here would be the job, the career, that would be my lifelong profession.

Being a publisher's rep was my dream come true, and I would have happily done that job for the rest of my days. You know the expression "Every cloud has a silver lining"? A friend of mine once reversed it. She said, "Every silver lining has a cloud." I am here to tell you that, backwards, it is every bit as true.

As luck would have it, back in the 1990s, independent bookstores had begun facing serious competition from the big chains. The mega-chains could offer customers lower prices than the independents around the corner, and soon these giants were just siphoning off customers. By the end of the decade, many independents had been forced to close their doors. You cannot blame folks for wanting lower prices, but it sure was and is sad to see those independent bookstores go. These stores, like mirrors, reflected their customers' tastes. They were community centers and the lifeblood of their readership. I have never walked into an independent bookstore whose staff could not tell me about the books they shelved in their stores. Talk about passion! I have spent many hours deep in conversation with booksellers, such as the famous Mary Gay

Shipley of That Bookstore in Blytheville, Arkansas, or Jean and Jerry Brace of Brace Books and More in Ponca City, Oklahoma, or Frances Comegys and Ginny Hill at Tower Book Shop in Shreveport, Louisiana, and J. Michael Kenney of the Book Merchant in Natchitoches, Louisiana. These book people are all literacy leaders in their communities. To me, these die-hard booksellers are national treasures, and they are my heroes for tirelessly crusading for literacy and overcoming all the adversity in bookselling today.

With each independent bookstore closing, I mourned that death as surely as a death in my own family. Faced with a shrinking account base, I doubled efforts in my remaining accounts, kept up my sales, and didn't worry. Heck, as long as I had *any*one to sell books to, I was a happy camper. Then, one fateful day in October, I was sitting at my desk, stuffing order forms into the spring 2000 catalogs, getting ready to hit the road for the accounts I would visit that week, when the phone rang.

"Good morning," I said brightly. My boss, Jack Richards, was on the line. I sat up a little straighter in my chair.

"Kathy," he began earnestly.

I wondered what was up. I knew that tone.

Jack called only when something was really, really wrong, or really, really right. I smiled to myself thinking, Maybe another bonus.

"As you know, Kathy, sales are down in your region as a result of all the store closings." He paused, as if expecting me to say something. If he was hoping I would get where he was headed, he was wrong. I did not have a clue.

"We've been forced to take a hard look at our operations . . . and . . . well, Kathy, it's the opinion of the partners that it is no longer cost effective to have a full-time rep covering your territory." He cleared his throat.

"We've got no choice, Kathy. We've decided to eliminate your position."

"You're *firing* me?" I asked, incredulous.

"As of today, that's right."

There was nothing but stunned silence on my end. For once in my life, I could not speak.

"Kathy? *Kathy?* Are you still there?"

If one big ole door slams in your face, a new window of opportunity will open. Give yourself wings and fly right in.

My head was spinning. This could not be

real. I was a good rep. I had kept my sales up. Sure, my numbers had not increased, but I thought I was doing a good job under the circumstances. I was a model employee, although I could not deny that, with only two years in the field, I ranked lowest in seniority.

"Yes, I'm here," I said weakly.

They say it is bad management style to fire an employee over the phone, but I was glad there was no one around right then to witness my reaction. I was trembling like a leaf; my stomach turned cartwheels. My mind flooded with nightmare fantasies: mountains of unpaid bills piling up on the kitchen table, creditors banging at the door. I silently screamed, Oh my God, what about our insurance premiums? The holidays are coming up! Not even to mention not one, but *two* car payments, which we had been carrying ever since I'd bought the damn minivan. I had needed it to haul around all the heavy boxes of books and catalogs for my sales work, a job that my previous car just was not equipped to do. My swirl of emotions included a quick burst of indignation — Wait a minute; I bought the damn minivan for this job and traded in my beautiful amethyst Intrepid! — which instantly gave way to helplessness.

We'll never make it on one salary, I thought. I felt sick. Vaguely aware that my boss was still talking, I came out of my trancelike state long enough to hear him thanking me for my top-notch service.

"It's not *you*, you understand, Kathy. You're a great rep, it's just —"

Right, I thought. I'll be sure to take *that* to the bank. (Funny how sarcasm will survive even the bitterest of blows.) I decided against an overt display of hostility. I muttered something in response, trying to meet his attempt at kindness with kindness of my own. Then the line went dead. So that was that. My dream job — my life — was over.

I don't know exactly how long I sat there frozen in time, not wanting to go forward, knowing I couldn't go back and recapture a past that was just a few seconds old. What felt like hours was probably just a few minutes. I looked down at the order forms and catalogs I had been preparing. Suddenly they looked foreign; they had nothing to do with me. I thought about finishing.

"Not my job anymore," I said to myself, scooping them up and tossing them onto a nearby chair. Then I picked up the phone to call Jay, my husband, to tell him that our world had just turned upside down.

As it turned out, the company agreed to

let me stay on another couple of weeks, so I could keep my scheduled appointments and take care of fulfilling orders for the all-important Christmas selling season. I said my good-byes to my beloved accounts and my colleagues, and called friends and family to let them know, trying not to reveal that I was devastated. It was my Elvis Presley moment; I felt just the way I felt the day the King died: inconsolable. After all, whenever anyone asked me how much I loved my job, I'd say, "How much do *you* love Elvis?" (That says it all to a Southerner.)

I was not exactly rolling in money, but we — Jay and my daughters, Helaina, nine, and Madeleine, five — got by. I had thought I would be selling books forever. Along the way, I would save enough money to be able to send my girls to the colleges of their dreams. That was the extent of my financial ambition. I had thought this job would never end.

"Now what am I going to do? What am I going to *do?"* I wondered aloud, over and over, until those nearest and dearest to me started wishing I'd get a grip. Sure, I could try to find more work as a rep, but no one was hiring. It was more like the opposite. Every rep I knew was hanging on to his or

her job for dear life.

I gave in to depression. Crawling into bed with books and my other favorite creature comfort, chocolate, I stayed there, neglecting my housework, my family, and myself. Jay wasn't exactly thrilled by this reaction. He would come into our bedroom and give me that "Now, Kathy, you know this isn't going to solve anything" look, which only made me dive deeper under the covers.

Though this is a cliché, come on, let's face it: I was seriously down in the dumps. You know the expression "mini-violins"? That was me all the way, having my own little pity party. Jay knew that any attempts to cheer me up — those things people say to reassure you like "Sometimes these things happen for the best" and "You've got so much to offer" and "You'll be just fine" — would just make me feel worse. Never mind that all those things are (usually) true and that the people are trying to be helpful. These were things I did not want to hear.

Jay was as worried about money as I was. I knew I would have to get another job. I just could not think about that yet. As far as I was concerned, I had hit the heights, career-wise. There was nowhere to go but down, down, down.

Jay picked up the slack, getting the girls

off to school in the morning, cooking dinner, helping with homework, and putting them to bed at night. Meanwhile, I buried my sorrow in reading *The Girls' Guide to Hunting and Fishing* by Melissa Bank and *The Nudist on the Late Shift* by Po Bronson under the covers. Books were the balm to my wound.

Then, one day, after I had read every book waiting in stacks on the nightstand and consumed I don't know how many boxes of Russell Stover chocolates, I woke up feeling completely fed up with myself and the whole melodramatic scene. Not wanting my life to go into diabetic shock, I put down the chocolates and did what I usually do in times of trouble. I reached for my "torch singer" Barbie phone on the nightstand and dialed my sister Karen. Two years younger in school than yours truly, Karen is the level-headed pragmatist of our family.

"What on earth am I going to *do,* Karen?" I moaned, not yet completely divested of self-pity and clearly without shame.

"Reopen your hair salon, Kathy," she said matter-of-factly, as if it were the most obvious thing in the world. She did not say it, but I *swear* she was thinking, *Duh.*

"Oh, Karen," I said in a pathetic voice. "I think I'd be so bored doing hair again."

She was unfazed by my whining. "Well, do the book thing, too."

I can honestly tell you that, right then and there, bells and whistles started ringing — no, clanging — in my head. To quote from one of my favorite children's books, "the light went on in the attic." A picture formed in my mind: a small beauty salon, with one, maybe two, stylist chairs and stations. Bookshelves stocked with only the best books, including all my favorites. To my knowledge, no one had ever tried this before, a hair salon/bookstore. I immediately saw the possibilities.

Looking back, I have to admit that this was one of the times in my life when things really did turn out for the best. Bookselling was intense, and I thrived on the challenge of helping a buyer select the perfect books for his or her customer base. But being a book rep involved a lot of travel, which meant days and sometimes weeks away from home. I would have to spend almost three weeks in the spring and three weeks in the fall at book publishers' sales conferences in New York City. I have never felt more alive than I did in my days in New York, but at the same time, I spent many sleepless nights tossing and turning, worried about being away from my girls. At that time, I was torn between career and home.

Sometimes when I wasn't on a business trip and I was working out of my home office, I would have to leave in the middle of the night to get to my accounts by nine in the morning. Kissing my girls good-bye as they were sleeping like angels usually made me bawl like a baby for the first hour or two on the road. I realized that as much as I loved that job, I also wanted and needed to be closer to home. I was soon to find out I could have the career of my dreams and still be a "good" mom.

Lainie and Madeleine were getting older. More and more, I felt they needed their mother at home. Now that I was away from the job, I could feel that tug — make that yank — on my heartstrings. I realized that if I worked closer to home, I could be there more for my family. Doing hair would give me a steady income that I could count on for cash flow. I could still talk books all day, and if a customer wanted to read a book I was praising to the high heavens, she could buy it right there in my shop. I could have my cake and eat it too.

Then it dawned on me: What about authors? I could invite authors to my store for book signings and do their hair and makeup first! I had helped with many author events in my years with books, and — no offense to any of my writer friends, past, present, or future — many of the authors I had seen just screamed, Makeover! I had a dozen more ideas, each one more exciting than the last. Suddenly, I was convinced that the world (well, Jefferson, Texas, at least) just had to have one beauty shop that sold books. Not only was I going to have my cake and eat it too, my cake was going to be frosted, big time!

> You can have your cake and eat it too, but remember: You have to buy the ingredients for the cake and bake it first.

"Wow," I said to Karen. "This is a fantastic idea. Why didn't *I* think of this? *This* could be my perfect job!"

In the blink of an eye I had gone from total despair to "Okay — when do we start!"

As I hung up the phone and threw back the covers, I jumped out of bed to hit the shower. I did not have to think about it twice. I was on a mission.

Many people questioned my sanity.

"It will never work, Kathy," they said. "Get yourself a real job."

Jay, who had majored in business finance

at the University of Texas, was one of the naysayers. Whenever the subject came up — which was often — he grew exasperated.

"Kathy, do you *know* how much money it takes to start a business, let alone a bookstore? Too much, that's how much."

Years earlier we had done a business plan for starting a bookstore, and financially the prospects of making a living selling books looked bleak. At that time I had abandoned the dream. Jay considered a hair salon/bookstore a financial impossibility. But now I was determined.

It's funny: I wasn't born rich. I may never see wealth in this lifetime, but when you look at it one way, I swear I act like someone born with a silver spoon in her mouth. You see, I believe I can do pretty much anything once I set my mind to it. I always remember that famous exchange between Ernest Hemingway and F. Scott Fitzgerald.

"The rich are different from you and me," says F. Scott.

"Yeah," answers Hemingway. "They have more money."

People will often repeat this when they want to tell you that money isn't the be-all and end-all of life. I agree: it's not. Though if you read between the lines, I think you can find in this exchange something a little

37

more life affirming. I think these statements also mean that when it comes to the things in life that are really important — family, friends, community, our health — we are, all of us, rich, poor, and in between, on a level playing field. If we set goals and make choices based on the things that truly matter, the sky is the limit on what we can achieve.

"No one can possibly achieve any real and lasting success or get rich in business by being a conformist."
— *Unknown*

Through good times and bad, I have held on to one belief. I believe that with the help of our faith, anyone can achieve anything. I am not saying be ridiculous about your dreams. You can't go dreaming of being an opera singer if you cannot carry a tune. But we can all take our passion and talents for something and follow those attributes where they lead. I suppose that is why so many of my favorite books teach us not to be afraid to dream.

Of course reality intrudes, even for optimists like me. To realize this particular dream of opening my own hair salon/bookstore, I needed capital. Capital was what I did not have at the moment. Luckily for me, it didn't take too much persuading to get Jay onboard. After just a few weeks of my constant wheedling, cajoling, and — let's call a spade a spade — *begging,* Jay took the $2,000 he had been saving to buy a golf cart and invested it in my crazy venture instead. Here was my knight in shining armor and what a princely deed indeed.

To give you an idea of the measure of his devotion, at the time Jay felt about golf the same way I feel about books. Almost every time I stop Jay as he is about to go out the door to tell him something, he stands there and practices golf swings. I can assure you I do not have his undivided attention. Just as when he is trying to tell me something, I usually have my nose stuck in a book. To give up his golf-cart money, my friends, that is true love.

If that was not enough, by November we had begun converting Jay's former office on the ground floor of our house into my new business adventure. He had given up his private retreat and moved his desk into the family room, sacrificing his domain for his fair maiden. My daddy then came down from Kansas to help me put in the shop, just as he had when I'd opened my first

If life hands you a great big lemon, forget making lemonade. I say grab a lime, invite over your girlfriends, and make some Marla-ritas!

Marla-ritas!

This recipe was given to me by my good friend and co-cosmetologist Marla Keith. You've heard the saying "It's five o'clock somewhere"? Marla-rita would say, "It's five o'clock somewhere — let's make some Marla-ritas!"

Ingredients:
1 ounce Cointreau
1 ounce Grand Marnier
1 ounce Jose Cuervo Gold tequila
2 ounces Jose Cuervo bottled margarita mix
1/4 lime
margarita salt
Mix all liquid ingredients in a glass pitcher. Take one large margarita glass and rub lime around the rim, rotate in margarita salt, add ice to glass, and pour in Marla-rita! Toss in the lime. Welcome to Marla-ritaville!

salon in Kansas.

I have a father who can do anything he sets his mind to, from fixing a car to plumbing for beauty-salon equipment. I don't remember an outside repairman ever fixing anything at our house or a mechanic ever working on our cars. The only other person who came to our house to fix something was my daddy's father. We called him Papa. He was an electrician by trade, and he always fixed our black-and-white Curtis Mathes television when it went on the blink.

When I called my father to tell him about my new hair salon/bookstore, I told him, "Daddy, I can't pay you anything, but I sure could use your help."

He was there the next day, and it's a nine-hour drive from Eureka, Kansas, to Jefferson, Texas.

> "You've got to sing sometimes like you don't need the money. Love sometimes like you'll never get hurt. You've got to dance, dance, dance, like nobody's watching. It's got to come from the heart if you want it to work."
> — *Glenda Jackson,*
> *British actress and politician*

I really got to know my father those

months of working side by side, getting the shop put together. I learned that my father may have had his faults, but he would do anything for us girls. He would work for us to his dying day, and he would give us his last dime. He may never tell me that he loves me to my face, but he told me that silently every day by getting up and going to work on my shop. He worked long and hard helping me get the shop together, and he is not a young man. If I had not lost my job, I probably would have not gotten to know my father in this way. Lookie here: that was just one of my wonderful windows of opportunity.

In January 2000, Beauty and the Book, the only combination beauty salon and bookstore in the country and maybe even in the world, opened its doors. Thanks to my King and Prince, Daddy and Jay, we were ready for business.

My crazy little venture succeeded beyond my wildest dreams. If someone had said to me back then that in five years I would move Beauty and the Book from my rural home to a historic house in downtown Jefferson, Texas, I would have said, "No way." If someone had told me that my book club, the Pulpwood Queens of East Texas, which started with six brave women, would grow to chapters running all across the United States and many foreign countries, I would have told him, "You are flat crazy." If someone had told me that I would work with companies like Redken and International Paper to promote literacy in communities throughout East Texas, that I would get to hang out with the writers who have for so long been my idols, and — the icing on my cake — that I would make an appearance on *Good Morning America* with Diane Sawyer and Charlie Gibson or see myself flashed on the screen during *The Oprah Winfrey Show,* I would have looked him straight in the eye and told him he was plum dee crazy. And yet, these and so many other wonderful things have happened since we opened Beauty and the Book.

Some days I still have to pinch myself to believe it's real. But it is real. Do I want to become famous? People ask me all the time, "Now that you've been on these television shows, how does it feel to be a celebrity?" I assure them that my upbringing has me securely grounded. Being on national television does not change your life — at least it did not for me. I still put my skirt on one leg at a time like everybody else, and guess who cleans the toilets both at home and at the shop? *Le moi,* the Pulpwood Queen herself, with toilet brush in hand!

Does it all mean that I am special? No sir-ree, Bob. All of this has come from having a deep faith in God and myself, the love of my family and friends, and hard work and determination. I took that great big old lemon that came from getting fired from a job I adored and just stirred up a big ole batch of margaritas — me and my Pulp-wood Queens girlfriends. We have been hav-ing a beauty of a book-loving party ever since.

> Throw yourself into good work and good things will come to you. Don't expect anything, and only then will the rewards for a job well done be yours.

The Pulpwood Queens' Tiara-Wearing, Book-Sharing Guide to Life isn't a book that will tell you the seven secrets of happiness. We all know that true happiness is fleeting. What my book will reveal is how one woman followed her passion and made her dreams come true. Mine is no Cinderella story, even though I do get to wear the tiara. I even went out and bought me some glass slip-pers, too. In my life, I have felt more like a Dickens character than Cinderella, if you want to know the truth.

In this book I will share the beliefs and values that are the foundation of everything I do. It seems that these values are in short supply today. I believe that, if I am able to live this dream of being "Hairdresser to the Authors" and book-group leader to many, then these values have had something to do with my success. We live in times where it seems like money — getting it, keeping it, and spending it — is the only thing people really care about.

I can tell you that I married for love, not money. I may be a Queen, but I was *never* a princess. You know the expression "Necessity is the mother of invention"? Isn't that what life is like for most of us? Like most women today, I have to work to help contribute to our family's day-to-day living. But I like working, and I like having my own money, and working gives me a feeling of self-worth. I figure that if I am working, I need to do something that my girls will see as important. You can talk until you are blue in the face about how you should do this or how you should do that. I just happen to believe that children learn more by your actions than by your words. I want them to see that you can pursue your passion and be rewarded, too, with things far greater than the almighty dollar.

There are four things I do believe helped me — five, if you count hardheadedness. The first is that I know that Jesus Christ is my Lord and Savior and that my having a spiritual belief in God will carry me through to life's final destination. Second, I have found my passion for reading and I let it be my lodestar. Third is having just enough faith in myself to get me through the hard times. The fourth is that I actually enjoy hard work and serving others. No one in this life gets something for nothing. It is far better to give than to receive.

> "We do not have a money problem in America. We have a values and priorities problem."
> — Marian Wright Edelman

My parents and my grandparents taught me that if you do a job right, you will be rewarded. I also watched them work hard at everything they did, and they made work seem fun. My grandfather Dirt always whistled while he worked on the farm. My grandmother Mudd sang as she worked in their shoe-repair shop and hoed the garden. My grandmother Murphy never smiled more than when she was sewing away at her Singer sewing machine. My other grandfather, Papa, always seemed happiest when he was tinkering on a television. My mother always sang when she did dishes and the housework. She taught us to dance while we helped with the housework. She would put on a Herb Alpert & the Tijuana Brass album or a stack of forty-fives and we would all dance away as we dusted and cleaned. My father believed that hard work would always provide for his family.

How do you go about finding that sort of satisfaction that comes about by doing valuable and enjoyable work? You can start by asking simple questions. What am I passionate about? Do I have a special talent or gift? What part of my life gives me the most personal satisfaction? The trouble with most of us women is that we are just so damn busy taking care of other people all day that we forget what we like or don't like. If we have talent, we downplay it. We hide our passion for everyday things such as taking care of children, mowing yards, cleaning house, or being a hairdresser. We think people will think less of us for enjoying those things. Let me tell you that there is absolutely nothing wrong with serving others. The sooner we realize that, the sooner we can give respect back to those who make our lives easier and more enjoyable. Caring for others is a calling, too.

Don't fall into the trap of thinking that success means having a big paycheck and the corner office, a big McMansion and a fancy car, or finding your fifteen minutes of fame. Every day since I opened Beauty and the Book has been a blessing. I may not have much money, but I am doing what I love, so I am rich in life. I've got plenty of riches — my faith, family, friends, my Pulpwood Queens, my books, and the deep satisfaction I get from promoting literacy in communities across the country. That's my story and I am sticking to it.

MY TOP FIFTEEN BOOKS OF ALL TIME

Tarzan of the Apes **by Edgar Rice Burroughs** Most books I discovered first by watching the film. I fell in love with the Tarzan movies and now own all of the books and collect all things Tarzan. Though I never wanted to be saved like Jane, I did want to swing through those vines like Tarzan and save everyone from evil. My childhood superhero was Tarzan, and you haven't lived until you have read one of his jungle adventures.

To Kill a Mockingbird **by Harper Lee**
Again, the film spoke to me in a way that I
had never felt before. I felt like I was Scout.
When I found out that *To Kill a Mockingbird*
is a book, I read it over and over, cover to
cover, now usually once or twice a year. I
can see no difference between the book and
the film. They are absolutely perfect, and
they are my standard for both books and
films today. Who would not want Gregory
Peck as their father? He read to Scout and
Jem every night before bed, and that is
heaven to me.

Before Women Had Wings **by Connie
May Fowler** There are so many lines in this
book that spoke directly to my heart that I
will treasure it always. The one that hooked
me from the get-go was "She named both
her girl children after birds, her logic being
that if we were named for something with
wings then maybe we'd be able to fly above
the shit in our lives." My sentiments exactly.

My Dog Skip **by Willie Morris** I have
loved so many pets in my life that just think-
ing about Skip makes me cry. Who wouldn't
love a book about a boy who loved his dog
and, more importantly, about the dog who
loved him unconditionally?

***Oldest Living Confederate Widow Tells
All*** **by Allan Gurganus** My grandmother

and grandfather Mudd and Dirt were story-tellers. I adored them, and those family stories helped shape me and still haunt me. Allan Gurganus wrote a story that captures you and has also haunted me to this day. As I read his pages, I was instantly brought back to a time when I listened to all my family's stories from Mudd and Dirt. You don't just read and hear the story; you experience it with all your senses.

***Rocket Boys* by Homer Hickam, Jr.** I always felt I was the underdog, and nothing appeals to me more than geeky unathletic boys winning full scholarships to college because they strived to do something different like winning the National Science Fair Competition. This is the book for anyone who thinks she can't when in fact she can.

***Crazy in Alabama* by Mark Childress** A woman who dreams of becoming a star in Hollywood told through the eyes of a child — slight difference in characters, but I lived this tale. I find it very reassuring in reading to find that I am not alone; there are others who are just like me.

***Charms for the Easy Life* by Kaye Gibbons** This is a story of three generations of women all living under the same roof. This book has such meaning for me, as my mother, my grandmother, and the other women in my family were and are always so important in my life. They had such an influence on me, and I would never be who I am today without these women. Kaye Gibbons is one of my favorite authors, and each of her books I hold dear.

Gone With the Wind by Margaret Mitchell My first viewing of the film *Gone With the Wind* was with my mother and sisters all piled on the divan with all the lights out so we could pretend we were really at the movies. I just loved Scarlett — when I read the book, I was able to recognize that we all make mistakes, sometimes big ones, but life does go on. She was my first woman hero.

Cat on a Hot Tin Roof by Tennessee Williams I have read everything written by Tennessee Williams, and I love the way he puts the fun in dysfunctional. I loved all the characters in this book. He taught me that all families have flaws; I always thought mine was the only one that wasn't perfect.

Little Altars Everywhere and Divine Secrets of the Ya-Ya Sisterhood by Rebecca Wells I feel as if I personally discovered this writer (I had read the advanced manuscript of her first book prior to publication), though in fact she was bound to become famous no matter what I ever

51

did or said. She has big talent and star charisma! Her books made me realize that we cannot choose our family but we must always love them, forgive them, and try as best we can not to repeat the mistakes of the past.

***Cold Comfort Farm* by Stella Gibbons** This 1930 British classic is one of the funniest books I have ever read. It's about a young woman who dreams of writing like Jane Austen. She decides in order to do so she must experience life first and goes to live with distant relatives in the country. This book may be set in England but reads as Southern as any Southern book I know, an absolute delight.

***The Great Santini* by Pat Conroy** To me, the greatest contemporary writer of my time is Pat Conroy. His prose is written with such honesty and longing, I despair on finishing his books. I love all his books, but the story of his father and family helped me to understand my own.

***The Rich Part of Life* by Jim Kokoris** This is the most beautiful and moving story about a man becoming a father. The author also has an uncanny ability to both be serious on one hand and amuse the reader to the point of laughing out loud on the other. You have got to love a book that moves you and cracks you up at the same time.

The Bible There is no other book I can turn to under any situation that keeps me from being a lost soul. Books have been my closest friends and companions. The Bible has been my saving grace.

You may wonder why there are fifteen books and not the customary ten or fifty or one hundred you usually find on "best of" lists. You may also wonder how a person with a lifelong habit of reading three or four books a week can limit herself to fifteen books. It's true that I have read and loved many more books than these, but if I listed all of them here I wouldn't have room in this book for anything else. I am sure you will see a pattern in the books I have selected. Yes, they are mostly Southern and mostly about dysfunctional families. These books are the ones that have touched me most deeply, the ones that I found, or that found me, at times in my life when I needed their wisdom and comfort most. I keep them close by and reread them often. And every time I do, I learn something new, see something I hadn't seen before, gain new insights into the human condition and into myself. They are my psychiatrist's couch. I could not have left a single one of these books out.

If you want to find out more about yourself, may I suggest, too, that you make a list of the books that touch you the most. Only you can decide their number, and what a wonderful gift you would have to pass on to others. I think the books we choose and love tell more about us than anything else. Passionate readers want to share, and this is a fact that ties my Pulpwood Queens bookgroup members together.

CHAPTER 2
DON'T JUDGE A
BOOK BY ITS COVER

(During a trial in which she was accused
of indecency on the stage)
JUDGE: Miss West, are you trying to
show contempt for this court?
MAE WEST: On the contrary, your Honor.
I was doin' my best to conceal it.
— *Mae West, American actress and*
sex symbol

People sometimes ask me what in the world
a bookstore and a beauty salon have in com-
mon. I look them straight in the eye and
state the obvious: Both are about friend-
ship, community, and feeling good about
yourself. They will walk into my shop and
turn around to leave, saying, "It's a beauty
shop." I will stop them and reply, "Wait, we
are so much more — we are a bookstore
too!" They stop in their tracks and turn
around as I explain what my shop really is.
Sometimes they will stay on for just a

couple of minutes; often they will stay for almost an hour. You see, they judged the book by the cover, when all they had to do was keep an open mind and experience what Beauty and the Book really is.

I believe that women go to the beauty shop for so much more than getting their hair done. The hour or two a woman spends each week or month at the beauty salon is often the only time she ever gets for herself, and for sure it's the only time she takes care of *her.* It is the same way with books. Most women think reading a good book is a luxury, what with taxiing the kids to and from piano lessons and softball games, picking up the dry cleaning, and all the other tasks women have to do daily. I, like many women, find it hard to indulge in anything without feeling guilty — except reading. Reading a good book helps us to escape from our lives for a while.

"While thought exists, words are alive and literature becomes an escape, not from, but into living."
— *Cyril Connolly, English critic and editor*

Have you ever read a book about someone

that you thought would read one way and then it turned out to be totally different? That is what happened when my friend and author Andy Behrman told me about Jeannette Walls's *The Glass Castle.* He told me a little bit about Jeannette, his good friend from college who was the MSNBC reporter for gossip and entertainment in New York. I had the idea that she was raised a privileged, educated New Yorker, a Manhattan princess. But when I read the book I found out that her family was from Appalachia and she spent most of her growing-up years homeless with her genius father and artistic, yet I think mentally ill, mother. I could not put that book down. I have learned that you should never judge a book by its cover; read it first.

My Beauty and the Book has been judged as a place only for women and, yes, it has become a safe haven for a wonderful community of women, but we also have men clientele. Women tend to be drawn to our shop, though, a world where they are comfortable talking, where they can just plop down and relax. Where they can drink a real Dr Pepper made with pure cane sugar and not be reprimanded for the calorie intake. We have a community spirit here at Beauty and the Book where we can let our hair down or put it up. We let women just be themselves.

I was inspired to be a Pulpwood Queen after reading *The Glass Castle* by Jeannette Walls. I never took the Pulpwood Queens seriously, and often laughed and even poked fun at their antics and appearance. However, as a media guest at the 2006 Girlfriend Weekend press conference, I was in awe when I saw MSNBC reporter Walls walking up the sidewalk at Kathy's bookstore/beauty shop, Beauty and the Book.

To realize that someone I admired as much as Walls would be a part of an event such as this changed my thinking and forced me to turn my feelings around and realize that the work of Kathy Patrick was worthwhile and not in vain. After that weekend I started a Pulpwood Queens chapter in my hometown and I have been wearing my tiara and reading ever since.
— *Phyllis, of the Pulpwood Queens of Marshall, Texas*

What happens at Beauty and the Book

stays at Beauty and the Book. I consider myself just like a doctor taking the oath never to divulge a patient's confidential information. People tend to tell their hairdressers everything, I think because of the magic of touch. The minute I start to massage and shampoo someone's head, I can see them relax. They let down their guard and talk about what is bothering them. I happen to believe that women need conversation, and we also need to know that information about us will not be shared with others. We do not gossip at Beauty and the Book. We share each other's lives.

The title for this chapter, "Don't Judge a Book by Its Cover," reminds me of the author Philip Gulley. Because Philip Gulley is a Quaker pastor and writes the Harmony series of books, which feature a Quaker pastor, I initially made some false assumptions. I assumed that Philip would be staid, dry, proper, and not anything like the wise, funny, down-to-earth fellow I have come to know him to be. Philip's books tell the tales of Sam Gardner, the pastor in a small town very much like my own Jefferson, Texas. Yes, I judged Philip by his cover when I really should have just read all his books.

Philip's publisher e-mailed me that he was just dying to come and visit the Pulpwood

Queens and evidently Philip got an e-mail from his publisher that the Pulpwood Queens were just dying to meet Philip Gulley. Whatever the reason, Philip Gulley came to visit us. Our lives were changed forever.

Philip and I got to know each other. When he asked me to describe Jefferson, I told him it was kind of like Mayberry on *The Andy Griffith Show.* We had our "Andy"s, our "Opie"s, and our "Aunt Bee"s, and I was the 2000 version of the barber, Floyd. Then I remembered Quakers did not watch television, or was that the Mennonites? Anyway, Philip set me straight and told me that when he was in seminary school, they had chapel every day. He found the hourlong service a tad boring and asked if there was an alternative. His superior told him that yes, there was: you could plan an independent Bible-study program. So Philip and his best friend at seminary did just that. He told me he was president of the Andy Griffith Theological Society. One half-hour they would watch *The Andy Griffith Show* in the church basement. Then they would have Bible study with scriptures that tied into the lessons of the show. I thought that was the coolest thing I had ever heard. I laughed my head off about it, actually.

I had no idea what a Quaker pastor believed, but I sure made some wrong assumptions. I figured he would be very strict, very stiff, and have absolutely no sense of humor. Boy, was I wrong. His books are hilarious, and I don't care what church you belong to or what your faith, you will find great truths in his books and some belly-hugging laughs.

When I wear my Pulpwood Queens attire (my "go to town" hair, hot pink T-shirt, black skirt, leopard shoes, acrylic nails, and lots of flashy jewelry), people sometimes assume that I am really "into beauty," that I'm a girlie girl. They might even say something like, "What were you, a beauty queen?"

I can assure you that nothing could be any farther from the truth. I am against anything that suggests that beauty is about how you look, because — it's a cliché to say it, but that's why it's true — beauty, *real* beauty, comes from within. What is important and what makes personal beauty is what is inside your heart, body, mind, and soul. How we look comes down through our parents' genes. We cannot help how we look (we can thank our parents for that), but we can help how we are on the inside.

> "Beauty is truth's smile when she beholds her own face in a perfect mirror."
> — *Rabindranath Tagore, Indian poet, playwright, and essayist*

I should know. My mother was a beauty. I thought she looked exactly like a cross between Jackie Kennedy and Marilyn Monroe, which is to say she had Jackie's dark hair and unusual eyes and Marilyn's curves. I grew up surrounded by the trappings of beauty and learned firsthand how it can be both a blessing and a curse.

People had always told her how pretty she was. Hear that often enough and you start investing in the idea, warming to the power it gives you. Men were drawn to my mother like bees to honey. I always felt that she relished the attention — in fact demanded it — and she always made the grand entrance. I always believed that she married my father because she said he was so good looking and he was so kind to children.

When we were little, my sisters and I were all towheaded children. We made such a striking contrast to both my parents' brunette hair coloring that people used to ask,

"Are they adopted?"

My mother would just smile politely and say, "Oh, no, these are my three *semi-beautiful* daughters."

If I heard it once, I've heard it a million times. Lucky for me, I wasn't quite sure what "semi-beautiful" meant. She always smiled so sweetly when she said it that I thought it was a good thing. Only when I was older did I begin to understand that being semi-beautiful was not quite the same as being beautiful. Years later, when I was in my early twenties, I asked my mother point-blank, "Mother, what do you mean by semi-beautiful?"

She matter-of-factly explained that it meant "almost beautiful." We just had not come out quite as good looking as our parents. The funniest thing is, all my life I thought I was not as pretty as my mother, but today, when people see my mother and me together, they say, "You look exactly like your mother." You can imagine my confusion.

I worshipped the ground my mother walked on. I believe now that my mother never felt she was good enough, pretty enough, and she set standards for herself that could never be met. She always expected failure, and because of that she did not want to set our hopes too high. I know now that she meant well. I do believe that she has always loved us in her peculiar way, and I know that we love her, unconditionally. We just sometimes have a funny way of showing it.

What I have learned from reading and also from looking back on my childhood all these years later is that we all really do the best we can. It seemed like my mother judged things and people by how they looked, simply because that was where her values lay. I know now that to see real beauty you must go beyond appearances and really try to get to know the person. I also know that people do judge you by the way you look, so I try to help everyone who comes to my shop put her or his best "face" forward. I help people look their best on the outside, but I also try to make people know that what is on the inside is most important. I like to compare this to a beautiful gift.

If the package has a pretty presentation but it's a disappointing gift, you lose. At the same time, if the present is wrapped haphazardly and the gift is exceptional, the wrap is forgiven. The best-case scenario is when the gift is presented beautifully and is given with much thought. Present yourself in the best possible light and have content of character. I like to think I help people achieve that presentation both inside and out.

I always thought we were poor growing up, but everybody else perceived us as rich. My mother gave us all a pretty fantastic and rich cover. We children may not have always had socks or underwear or clothes to wear to school, but my mother drove a brand-new Cadillac. Our cupboards might have been bare, but we belonged to the country club. My mother would stay at home for weeks, but when she went out, she would drive in style like a queen. I often felt like Cinderella in rags and ashes, yet we lived in a castle. To me, my mother was oftentimes like the evil stepmother. She would leave our house all decked out in the finest clothes. My sisters and I were left to fend from her castoffs and patch our own things. She told me later that it really did not matter what the children wore because people judged you by your parents. I think this was fairly true of her generation. Parents dressed up when they went out or went to work. Children were to be seen — briefly — and not heard.

My mother's beauty ritual was a staple of my childhood, and the routine never varied.

It always began with a long, luxurious bath. She would turn up the taps to steaming hot, then sprinkle in a few drops of Estée Lauder Youth-Dew bath oil. While the tub filled, she took a towel and wrapped it around her head like a turban. I had seen this done in magazine ads or on TV, but it was a feat I could never duplicate. Whenever I tried to do it myself my towel would fall into the water and my hair would get soaking wet. I was fascinated by her skill and dexterity — my mother's turban never came undone, naturally. Then she would step into the tub and soak for at least one half hour.

Next she would shave her legs, and she did this every day whether they needed it or not. Then came my favorite part. When she was finished with her bath, she would wrap herself in a big white terrycloth towel and slather her body with Jergens hand lotion.

I would watch her drop her towel and walk stark naked from the bathroom to her bedroom and slip on a see-through baby-doll nightgown that was usually a pale pastel tone of blush or twilight. She was never embarrassed by her nakedness; in fact, she seemed very comfortable in her skin. (God forbid anybody should ever see me naked! I would rather be strapped to a stake and be burned alive than be seen naked by anyone.)

"The best mirror is an old friend."
— *George Herbert*

She would go back to the bathroom and sit at her built-in dressing table with counter-to-ceiling mirrors. There, looking at herself the whole time, she would begin to set her hair with big brush rollers, which she held in place with pink picks and bobby pins. Then she would parade back to the bedroom and take out her Sears Sunbeam cream-colored portable hair dryer and carefully stretch the plastic bouffant cap over the curlers so as not to knock one out of place. The fancy shower cap had a hose that was connected to the blow-dryer that, when switched on, would blow hot air into the cap to dry her hair. She had propped pillows up against the bookshelf headboard of her blond Hollywood bed. The shelf might hold one or two *Reader's Digest* condensed books, but it was mostly filled with her beauty products and prescription eyeglasses, which she never, ever wore in public. In her opinion, it was better to be blind than wear glasses.

While she waited for her hair to dry, she

gave herself a pedicure and polish with a Revlon frosted pale pink tone. Color on your fingernails and toenails was tacky. God forbid you ever wore hot pink or red. That would have been absolutely sinful. Fingernails were manicured last with the same pale pink frosted polish. She was never in a hurry as she blew on her nails and seemed perfectly content just to sit, or preen, or maybe flip through a fashion magazine she had picked up at the grocery store, until her hair and nails were dry. This process seemed to drag on for hours.

After her hair was dry, she would take off the cap, put everything away, cleaning as she went, and then go back to her dressing table. Next came makeup, perhaps the most amazing transformation of all. Dabbing her forehead, cheeks, and chin with drops of her Helena Rubinstein foundation, she would blend her skin using little swirling motions, stopping occasionally to study her reflection and measure her progress, until it was completely smooth and even. (I remember once sneaking into her room when she wasn't there to try it myself. On my "virgin" skin it appeared very dark. My mother's complexion was olive and I had the skin of the Scot/Irish: pale, susceptible to the sun, and prone to freckles. Not yet understanding the desire nor feeling the need for artifice, I had zero tolerance for its discomfort. I could not wait to wash it off.)

Then from her makeup drawer came this little red box of Maybelline mascara, the "cake" kind that you had to moisten yourself, which she did by spitting into the cracked black crème. She would carefully brush it onto her eyelashes, making sure it did not clump. Then out came the eyelash curler, a metal contraption that looked like it could do serious harm to a naked eyeball with just one slip of the wrist.

> "The face you have at age twenty-five is the face God gave you, but the face you have after fifty is the one you earned."
> — *Cindy Crawford*

"Be very careful when you lift the curler off your eyelashes," she would warn, no doubt hoping I'd be repeating her beauty ritual one day. "I had a friend who took off all her eyelashes that way and they never grew back."

Her eyebrows were plucked to perfection, and she had one brow that, when cocked, had more of an arch than the other. I have my mother's eyebrows, though I never could cock one just like her. She would pencil them in and then brush on a slightly darker brow color, never in a hurry, as if there were all the time in the world.

Of course her lipstick was frosted pink (to match her nail polish). She applied it with a little gold, retractable lip brush. She finished with a light dusting of powder and then it was time to get dressed.

> There is nothing more beautiful on someone's face than the look of true happiness. Happiness nor beauty comes from a jar; they come from within.

Her closet was a virtual smorgasbord, a fashion buffet fit for any occasion. From bridge-club ensemble to after-five formal, each outfit had its own matching jewelry, matching shoes, and matching handbag, and, if the occasion called for it, a matching hat. My mother believed in matching entire ensembles perfectly, à la Jackie Kennedy.

Her walk-in closet ran the full length of the upstairs, with shelves down one side for hats, shoes, and purses, all in their original boxes and labeled for quick reference. The other side of the closet had one long rack where she hung everything, color-coordinated by outfit and by season. The blond dresser in her bedroom had drawers for gloves, undergarments, scarves, and sweaters, each carefully wrapped in tissue with drawer sachets. Some of the panties, slips, and scarves would still be in the slender black and beribboned boxes. We were not allowed to open her drawers or touch anything, and of course that's exactly what we did the minute she was out of the house. We would carefully open and peek into each package and be extra careful putting everything back just the way it came out.

I share the philosophy of Desiderius Erasmus, the Dutch priest, who said, "When I get a little money I buy books, and if any is left I buy food and clothes." My mother would rather buy food and clothes than eat. I never remember her sitting down to eat with us except on the rare holiday occasion.

She had a blond side table with a drawer filled with even racier nightgown and negligée sets. Underneath all the lingerie would be one or two *Playboy*s that we loved to sneak out to check out the cartoons. The cartoons with the big bubble-headed blonde with the big blue eyes and gigantic boobs always made me mad. She was so stupid, letting men get away with all the things they did to her. We thought it was creepy that the women were always naked but the men never were, but then we remembered that our mother always paraded around naked and our daddy never did.

My mother would lay her selections for the day on the bed, then proceed to assemble a battery of undergarments, weapons in our mother's war against gravity and other natural forces. First came a bra that resembled nothing so much as a harness. She had undergarments in every available hue of white, beige, and black, some even completely see-through, sometimes with lace. Then came that true artifact of prefeminist America: the full-body girdle. They were designed to make sure a woman's curves were in the right places while quieting any jiggling flesh. These excruciatingly uncomfortable pieces of "armor" (is there any other way to think of them?) served one practical function: to make everything appear as taut and firm as possible. The entire look was to be as tight as a drum.

My mother always wore garters to hold up her thigh-high hosiery; these foundations she wore no matter how humid or hot the day might be. Now, it's a scientific fact that synthetic fabrics don't breathe. If you are wearing nylon stockings on a typical blazing-hot summer day in Kansas, you are surely going to sweat. Sweating, as far as my mother was concerned, was the height of female uncouthness.

"A lady *never* sweats, Kathy," she often told me. "It's crude."

So out came the next weapon in the arsenal: dress shields. These were big, thin cotton pads that fit directly under your armpits and were held in place by white straps. The whole thing fit over your bra. I thought they were about the silliest-looking contraptions I had ever seen, and I vowed never to wear any of those torture devices. No matter how entranced I was by my mother's beauty ritual, I would rather be a tomboy if that was what it took to be a woman. Whew!

When we were very little, in our pre-school days, my mother would try to get me to wear starched and ironed dresses with built-in crinolines and a big bow tied in the back. The collar was always too tight on my neck, the petticoat itched, and that knot from the bow in the back of my dress always killed me when I sat back in my chair at school. I can remember taking the worst ones and hiding them behind an old divan out on the screened-in back porch.

"How could you lose a dress when it was practically brand-new?" she would lament as she searched through our closets and chests of drawers for the dress I had hidden behind the couch.

By the time we started school and started to dress ourselves, really rebelling against wearing anything uncomfortable and itchy, she gave up in mock despair. Besides, we had lost our baby cuteness — especially after Mother decided to cut our rapidly growing, long blond tresses in the latest razor-cut pixie style. Only she happened to use scissors, and the effect was not quite the same. Our heads looked as if they had been hit by a lawn mower. She just kept cutting and chopping hair, trying to get it even. She would dip her comb in a glass of water to wet our hair. She never could get bangs or sides straight, so she just kept whittling away until we hardly had any hair left.

Looking back, my sisters and I all looked like children from the Holocaust with our too-skinny legs under too-short and too-small dresses with our crooked and ragged haircuts.

During my teens, my basic disinterest in appearances morphed into wholesale rejection. For years I did everything possible to be the complete opposite of my mother. If she dyed her mousy brown hair jet black, set it on rollers, backcombed and brushed it out to create a little Jackie-like flip, I wore mine long, blond, straight, and parted down the middle. If she favored dresses that showed off her hourglass figure, my "uniform" became blue jeans worn with work shirts, which, it goes without saying, I didn't tuck in.

I have always hated belts. My mother simply could not understand this. We used to bicker about it all the time, with me marshaling all my powers of persuasion, which she stubbornly (and stupidly as far as I was concerned) resisted. One day I was sure I had found the ultimate zinger.

"What about the ancient Chinese practice of foot binding?" I cried, triumphant (Aha! Take *that!*). "Those poor women may have achieved the ultimate pinnacle of beauty — *but they were permanently crippled!* No one is going to catch *me* squishing up all my internal organs just for appearance's sake."

"Kathy Louise Murphy," — she harrumphed — "*I* have been cinching in my waist ever since I was old enough to *have* one." Her tiny, twenty-one-inch waist was her constant source of pride. "It never did *me* any harm, and it wouldn't hurt your waistline a bit either. In fact, a man could circle my entire waist with his hands."

Neither my mother nor I ever "won" the argument, and "to wear, or not wear" a belt, that was the question. To this day, she swears that the only reason I don't have a waistline is that I never wore belts.

> "A waist is a terrible thing for the mind."
> — *Jane Caminos*

I don't think my mother meant to be cruel. At that time a lot of women wore constricting garments. My mother just took it to the extreme. For her, it seemed that beauty *was* all about how you looked. She seemed a slave to her own beauty in many ways, and I am sorry to say it left room for little else.

Still, no matter how much we reject the past, it shapes us. I think of all of the professions I could have picked in the world. I chose to run a beauty shop. How ironic. I see it like that cigarette commercial they used to show on television when I was a child. A man is sitting down by a tree to have a smoke. On the other side of him his little boy sits down and mimics the whole thing. Yes, actions do speak louder than words, and I have now made a profession out of the focus of my mother's daily ritual. But I can tell you that my shop is not your typical "beauty" salon. As I have said, true beauty comes from within. In my salon, I am just helping others feel good about themselves, not creating a false persona.

To me, beauty is like a mirror — not one you should stare into for hours, preening and flaunting, but a mirror that reflects what others see. The evil Queen in *Snow White* let her shallow desire for perfection be her downfall when we could all see that no matter how beautiful you are on the outside, if you are not beautiful within, you are ugly. Snow White's love for others and her inner beauty far surpassed the Queen's superficial beauty.

> "I used to be Snow White, but I drifted."
> — *Mae West, American actress and sex symbol*

These mirror reflections should be a vision that makes you feel good about yourself. How many times in life have you met someone who, at first, did not seem very attractive, but the more you got to know him or her, the more attractive the person became? The beauty in people's hearts and souls, their inner beauty, is what you see. When we radiate love and compassion, beauty shines through from within us. There are many lessons we can take from the childhood fairy-tale books of our youth.

You can spend hours in front of a mirror trying to make yourself look beautiful, but it will not do you a bit of good if, inside, you are thinking negative thoughts. I try to be that positive mirror for my clients. I cannot tell you how often a customer comes into Beauty and the Book for her first appointment and hands me a picture of Jennifer Aniston or Paris Hilton or whoever is the "flavor of the day" and says, "Can you make me look like that?"

My first question is not what they expect. "Why?" I ask.

Nine times out of ten they can't answer me. So I tell them, "First off, I could give you a Jen cut, but you still wouldn't look exactly like her. Your hair would look like you with a cut like Jennifer Aniston's. Only Jennifer looks like Jennifer; only you should look like you. Why would you even want to look like a carbon copy of someone else! I would much rather create a look that is totally you and one that enhances the best of your features."

My clients always smile when I say that. I then explain why that particular cut would or would not be the most flattering for them based on their face shape. Maybe the texture of their hair is thick, wiry, and curly, or so fine that you can practically count the hairs on their head. They really listen to me, and I try to really listen to them. That first consultation can take some time, but I like to give myself plenty of time to find out my client's life-style and exactly how much time she invests in getting ready. Usually I find that she truly does not know what would look good on her, so I steer her into what would accentuate her positive features and stay away from the negative. Too many times we focus on being too fat, too thin, having a big nose, being too small in the bust, having hips that are too wide, legs that are too skinny. Everyone has something that makes them unique and beautiful. My job is not to focus on what society deems beautiful, but to find the one thing that makes each woman unique and — yes — beautiful.

When I have someone new come into my salon, I try to put myself in her shoes. I never do anything to any client that I would not do to myself, and I really try to treat people as I would want to be treated. I find out their likes and dislikes, hopes and dreams. I cannot do a beauty makeover until I know who my client is as a person. I want my customers to see and value their uniqueness.

If you want to find a good colorist and hairstylist, look to others who sport something you like. Ask them where they get their hair done and book an appointment with that stylist for just a shampoo and blowout. Ask the stylist what she would do to your hair based on your appearance and lifestyle. If he or she cannot answer you, do the scheduled blowout and keep looking elsewhere. If he or she can tell you and you like what you hear, book another appointment with that stylist for your next salon service.

Can you imagine Cher or Barbra Streisand without her nose? Or Cindy Crawford without that mole? Or Lauren Hutton without that space between her two front teeth? I happen to believe those things are what give us character. We all can think of certain celebrities who tried to change their appearance through the various cosmetic enhancements and plastic surgeries and everything went wrong. Don't get me wrong — I find no fault with having those things done by reputable people for all the right reasons. But I happen to believe that good skin care, good grooming, tasteful clothing, proper hygiene, and good manners impress people more than Botoxed wrinkles, eye and boob jobs, and liposuction.

> "I only answer to two people: myself and God."
>
> — *Cher*

People who are unhappy with their looks are looking inward instead of outward. I understand, too, that physical deformities can crush a person's spirit, and that is exactly why we should have cosmetic surgeons. But today it's not uncommon for young girls to get boob jobs for graduation gifts! I'm sorry, but that is just taking vanity much too far. I have also found that most people really don't give a rip what you look like, as they are just too busy worrying about how they look, or thinking that what they may be wearing is wrong.

We may not always be able to be perfect on the outside, but we make up for it by working on what is in the inside. When you start getting all whiny about how you look, I believe that you should pick up a good book. Reading can take your mind off yourself and give you the insight and direction to realize that it is not all about *me* or *you!*

I have a client named Leslie who wanted to exchange ad space in her magazine for her hair care. She had had some coloring done, and it had grown out and the shades were not to her advantage. I added some highlights and a lot of low lights. She gave me complete license to change her look. When I finished she was amazed. She stood up, walked closer to the mirror, and just screamed, "I love my hair!"

I told her this look was original, just for her. Since then, Leslie has come back, and she leaned over and told another client as I was mixing up her colors, "This is only the second time I have been to Kathy's shop, but I will travel to the ends of the earth as she makes me look and feel so special."

She did, and I have never been happier.

Another client once pulled me close to her and told me, "Kathy, I have been to a lot of hairstylists and some who may have done my hair better. But I keep coming back because when I leave here, I feel like a million bucks."

Well, I'm not sure about other stylists doing her hair better, but I sure liked the million-bucks part!

The most amazing client, my friend Dona, told me and has told me over and over again, "I love coming to you, Kathy, as I am a very shy person. You never make me feel self-conscious but make me feel like I am loved. I can live vicariously through your life. Through your stories and books, I feel I have done all of those things. Besides, I love what you do to my hair."

I can die a happy woman when I hear that kind of beauty.

Unlike my mother, I have a beauty routine that takes about twenty minutes, start to finish, and that includes my shower! I use a tinted moisturizer, a light brown pencil to define my brows and frame the rest of my features, a little blush, mascara, and lipstick. I'm ready to go. And that's on work days!

On my days off, it's no makeup, just good skin care and a baseball cap over a ponytail for me. It just slays me to hear women say they never leave their house without makeup. My mother was one of those. We called it "putting on her face." Putting on a mask is more like it.

If you ever start thinking you don't look good unless you're wearing makeup, that's

> Be happy with who you are. Beauty does not come in a perfect package. Remember the first homemade present your child ever made you, or you made for your parents: precious but not perfect, not even close.

a sign of a dangerous "makeup dependency." My advice is to go without it for a day, a week, or a month — however long it takes you to realize that your friends still recognize you and quite a few people won't even notice. Makeup should *never* be a mask.

There is great truth and beauty in books. I found that Philip Gulley's books and the following are ones that spoke to me and made me realize that true beauty does come from "within" — within a good book, of course! These books showed me that there were people who had worse circumstances than mine. My mother was obsessed with beauty. I am obsessed with creating a new kind of overall beauty that celebrates the intellect and physical beauty at the same time. To my mother's credit, she did subscribe for us girls to receive the Golden

Book Encyclopedia volumes and books on different countries that came with stickers. We read those books cover to cover. I like to think of it as melding my mother's values with my own. These are books that made me look outward instead of in. I hope they reveal these qualities to you.

***Crazy in Alabama* by Mark Childress** How can you not love a book about a Southern woman who kills off her redneck husband, dumps her kids at her mother's, and heads to Hollywood to star on *The Beverly Hillbillies*? Sounds crazy? You bet it is, and Mark ha9s the ability as a writer to take something totally whack and make it completely believable. This is black comedy at its finest, told by the young nephew, Pee-joe, during civil unrest in the South in the 1960s.

***The Liars' Club* by Mary Karr** This memoir of growing up in an East Texas refinery town made me realize that there is great truth and beauty even when you are being raised by liars and drunks.

***Bastard out of Carolina* by Dorothy Allison** A National Book Award nominee. I learned from this book that you cannot choose your family, and that both reading and writing can help you recover.

***Touched* by Carolyn Haines** Dirt-poor sixteen-year-old Mattie is married off to the town's barber and becomes friends with the small Southern town's most scandalous woman, Johanna. When Johanna's daughter gets struck by lightning and lives to foretell the future, all hell breaks loose. Crazy Southern stories are something I can relate to, and this one really made me sit up and pay attention.

***All Over but the Shoutin'* by Rick Bragg** Rick Bragg's love letter to his mother made me understand that your mother is one of the most important people of your life, and that you should never forget your roots.

***If Nights Could Talk* by Marsha Recknagel** This family memoir is the story of a single Rice University writing professor whose sixteen-year-old nephew shows up on her porch crying, "Help, save me." He had run away from his parents (who met in a mental institution); the story also involves a six-year battle for the family oil fortune. It's so crazy you could not make this story up.

***Screen Door Jesus & Other Stories* by Christopher Cook** Set in the East Texas town of Bethlehem, this short-story collection asks not whether you believe in God, but how you believe. In the title story a woman discovers an impression of Jesus in her screen door; it is based on a true occurrence.

***The Secret Life of Bees* by Sue Monk Kidd** A young girl is on the run with her father's servant searching for a mother. They are taken in by three women who are beekeepers but make more than honey. A tragic accident made young Lily lose her own mother, and one of the women teaches her how to find the "mother" in herself. I found this very life affirming.

***Why I'm Like This: True Stories* by Cynthia Kaplan** A funny collection of essays where the author is trying to find herself and comes full circle.

***A Hole in the World* by Richard Rhodes** There are a lot of tell-all memoirs out there these days, but this book goes into a whole new dimension. The author's mother committed suicide when he was thirteen months old, thus setting forth a life that is worth reading about. What will impress you the most is the overall tenacity of the human spirit.

***My Brother Michael* by Janis Owens** A haunting Southern family saga with a lot of guarded secrets. Another Southern gothic that has a flawed protagonist who finally receives redemption.

***Blackbird: A Childhood Lost and Found* by Jennifer Lauck** When the author was five years old she was responsible for taking care of her dying mother. Shortly thereafter, her father moves her and her brother to California to move in with his girlfriend, Deb. Deb already has a ton of kids, and the new daughter is eventually forced out of the house. The picture of the author as a young girl dragging her white princess bedroom set down the street to a halfway house will be forever ingrained in my mind.

***Still Waters* by Jennifer Lauck** The continuing memoir of Jennifer Lauck, which begins with her grandfather picking her up at the bus station. You mistakenly think she is saved, but she soon moves in with her aunt and uncle. After reading this book, I swore I would never feel sorry for myself again.

***Once in a House on Fire* by Andrea Ashworth** The author of this memoir holds degrees from Oxford and Yale. By reading her book I have learned that children from dysfunctional families do overcome great adversity.

Want to know the best way to make someone feel beautiful? Give a person your sole and undivided attention by making eye contact and really listening to what he or she has to say.

CHAPTER 3
THERE'S NO PLACE
LIKE HOME

"The Road to the City of Emeralds is
paved with yellow brick."
— *L. Frank Baum, author of*
The Wonderful Wizard of Oz

I grew up with a vivid imagination, which
may be hard to believe of someone who is
from Kansas. Then again, remember that
Kansas is where *The Wonderful Wizard of Oz*
was set.

Most people think that Kansas is as flat as
a pancake, as it was portrayed in the film
The Wizard of Oz. My hometown, Eureka,
surrounded by rolling hills, prairie grasses,
cottonwood trees, and meandering creeks,
is a far cry from the land around Auntie
Em's farm. To a child growing up in the
1950s and 1960s, Kansas could be a magi-
cal place where kids could run wild. From
the first day of spring until school started
again in September, my sisters and I practi-

cally lived outdoors.

Just like Dorothy in *The Wizard of Oz,* I longed for someplace to go to escape. "Somewhere over the rainbow" for me was more than likely just outside. I loved to escape to our tree house, where at least I was way up high. I have seen *The Wizard of Oz* as many times as years I am old (at least fifty times) and have read all the books. I never miss an opportunity to see a production of *The Wizard of Oz.* That book and the film mean as much to me about home as home itself.

When I think of Oz, I see the yellow brick road leading to the fantastical Emerald City. We may not have had the Scarecrow, the Lion, or the Tin Man, but we did have our own Toto, our little Jack Russell–looking mutts, first Pepper then Snicklefritz. We spent hours pretending to be caught up in a tornado. The wind always blows in Kansas, and with just a little imagination my sisters and our friends and I were on our way to Oz!

Some of my fondest memories are of building our tree house, flying kites and making mud pies, playing kick-the-can or hide-and-seek until it was pitch-black outside, then catching fireflies in mason jars. We created our own worlds with our own imaginations. I believe that our imaginations and our dreams are as important as our basic needs: food, water, and shelter. Many times those dreams were all that sustained us during life's trials and tribulations.

> "Dreams are the answers to questions that we haven't yet figured out how to ask."
>
> — *Fox Mulder*

To keep ourselves in candy money, we might set up a neighborhood lemonade stand, raiding the kitchen cabinets for empty grape jelly jars featuring pictures of the Flintstones or Yogi Bear, the kind that had a second life as drinking glasses after the jelly was gone. Another favorite activity was "creeking." There wasn't much to it. We'd walk through the big pasture behind the nursing home across the street from our house until we got to the tree-lined creek. In our imaginations we were pioneers crossing the great Midwestern plains in a wagon train, stopping to wash up in the first fresh water we had seen for weeks. Or we were outlaws trying to outrun the sheriff and his posse by throwing them off our scent in the rushing river currents — which were more like a trickle, but I had a big imagination in those days. We would go as far as we wanted, not stopping until the sun began to sink on the horizon. Only then would we hightail it back home, muddy and dripping wet.

As long as we *were* home before dark, our mothers didn't mind. (Auntie Em never worried about Dorothy on the farm. Well, not until the tornado came blowing in.) Shoot, nobody's mother gave us a never mind. Just as long as we were out of the house, the mothers in our neighborhood were happy. To tell the truth, while our dads were at work — ours was gone sometimes for weeks working on the drilling rigs — our mothers were too busy with housework to want us underfoot anyway. Besides, half of them had grown up doing these same things themselves. They knew the mischief we might get into and they knew its limits.

Today the world is a more complicated place. I can't imagine giving my girls that much freedom. There are too many wicked witches and flying monkeys out there in our world. In fact, just thinking about my girls doing half the stuff we did as kids scares me to death. I am more like the overly protective farm hands who fussed over Dorothy when she awoke from the bump on the head.

I realize that some might find me sentimental as I recall these simple pleasures. I don't deny it. Like most childhoods, mine contains happy memories as well as some that aren't so happy. When I look back, I remember the happy times first.

> Everything in life is perceived by your attitude and mind-set. I have learned if you think happy thoughts, you will be happy.

I love my parents dearly and wish them all the happiness in the world, but to say I had the perfect childhood would be an out-and-out lie. As independent and free as I felt outdoors, it was another story inside the house. Inside, I tried to be quiet and obedient, hoping it would keep me out of trouble. My first memory is playing on the back cement steps of a white two-story house we lived in on Illinois Street. I must have been two or three at the time. I was squishing my toes in the cool, silty mud and it felt really good. I was also making an assortment of mud balls, which I had lined up in front of me on the bottom step. One by one, I would roll each ball in the palm of my hand, turning the smaller ones into worms and the bigger ones into snakes. I was totally engrossed in my project when I heard the screen door squeak open above my head. My mother's raised voice shattered the silence.

"Kathy Louise Murphy! What have you done? Just look at that pretty dress I just ironed! It's *filthy!* You get in this house right this instant!"

I was too bewildered and terrified to disobey, but she was not taking any chances. She reached down, grabbed my arm, and pulled me inside. Like Dorothy, I felt trapped and doomed by the Wicked Witch of the West. Though when my mother was all dressed up and going to town, she was just like Glinda the Good Witch. The wonderful world of Oz was a confusing place, as was my home. I never knew if Glinda or the Wicked Witch of the West would be waiting for me when I went in the door. My sisters and I laugh about it now as we recall that when our mother got angry she would slap us and then grab our shoulders and shake us. Of course, this kind of discipline seemed commonplace back then, unlike now. In fact, the Wicked Witch of the West is my favorite character in the book, and I have the movie version down flat. My girls used to scream with glee, "Do it again. Do it again, Mama. Do the witch!" Then I would go into my best Wicked Witch of the West impersonation. "I'll get you, my pretty, and your little dog, too." As teenagers they still sometimes ask me to do it for their friends. I have no idea why we find it so funny, but it cracks me up every time. Maybe my reasoning got shaken a bit loose by my mother shaking me as a child.

> "For twenty-three years I've been dying to tell you what I thought of you! And now . . . well, being a Christian woman, I can't say it!"
> — *Auntie Em from the film*
> The Wizard of Oz

If my mother had on her black cat-eyed glasses and wasn't wearing her makeup, it was always worse. The truth was that she was always nicer when she had her makeup on. Most of the time she just scared me half to death, just like my father. My sisters and I both recall being scared a lot of the time. We would all crawl in together in bed at night — that was when we were scared the worst.

Everyone has a story, and this is what I know about my mother. Mary Louise Maloney was a local beauty with dark hair, hazel eyes, an olive complexion, and a voluptuous hourglass figure. Are looks destiny? Sometimes I think so. She had wanted to be an actress, and for as long as I can remember she would tell stories to my sisters and me. She had great stories.

When my mother graduated from Hamilton High School, a small rural school not too far from Eureka, she enrolled in Emporia State Teachers College as a speech and drama major. She was seventeen. During her sophomore year, she always told us, one of her professors, a Dr. Phfalm, thought she had exceptional talent and recommended that she try her luck in Hollywood. If she were interested, he said, he was pretty sure he could find her an agent. It took my mother all of two seconds to say, "Goodbye, Kansas. California, here I come." I am sure when she got there she was kind of like Dorothy saying, "Toto, we're not in Kansas anymore."

Upon her arrival in Hollywood, I can more than imagine my mother felt she had landed in the wonderful world of Oz. Only if you looked on the other side of the set, you found unfinished frames and boards.

Fame and fortune weren't in store for my mother, but she did come home with a suitcase of memories jam-packed with the Hollywood stories that filled our childhood. I especially remember one about my mother and her agent having dinner with Agnes Moorehead, who later became famous on the television show *Bewitched.* Mother explained it was the thing to go out and be seen at a really ritzy restaurant with someone who was already famous. Whenever Mother told the story, she exclaimed, "Ms. Moorehead had the most beautiful skin I've ever seen, just flawless."

Agnes Moorehead played Endora, the mother of the main character, Samantha. They were both witches, and my sisters and I endlessly practiced trying to get our noses to twitch back and forth like Samantha's did on the show when she was casting a spell. We always found it hard to believe that Agnes Moorehead had such a flawless complexion because on the show Endora wore so much atrocious makeup, especially that turquoise eye shadow and black eyeliner. It was hard to imagine how her skin really looked.

Mother, unfortunately, did not get discovered in Hollywood. But she still always had stars in her eyes and dreams of Hollywood.

On special occasions she would bring out homemade paper dolls she had drawn and colored as a child, mimicking the 1940s style and glamour of her *Modern Screen* and other movie-star magazines. The cartoon character Brenda Starr, with stars drawn in her eyes, was a favorite rendition in her paper dolls. We could look, but we were not allowed to actually touch them.

My mother lived for the movies, and we girls too were caught up in all the trappings and glamour of Hollywood. I thought all people in Hollywood wore beaded gowns, had mink stoles, and rode in high style in limousines. Even as small children we went to the movies a lot, at either the Princess Theatre downtown or its partner business, the Highway 54 Drive-In. We watched the *Sunday Big Show* on television and, if a movie came on the lights went out. Every spring we would pile onto the divan in our pj's, glued to the black-and-white picture, all oohing and ahhhing when Dorothy landed in Oz and the image changed to color. That is pretty much the only time I remember my mother actually holding my hand or cuddling close to me or my sisters. My mother was always happiest when she was lost in the magic of the screen.

Every year it was a ritual to watch both

The Wizard of Oz and *The Academy Awards.* Mother would pop popcorn, we would have real Cokes in tall glass bottles, and lemons cut in half that we would sprinkle with the saltshaker and suck on during the show. To this day, I never miss those two shows when they air on television. I gather my girls just like my mother did, on our leather couch with popcorn, Cokes, and lemons. It's a family tradition. These times remind me of the best of my mother, and I want my girls to be reminded of the best of me.

Mother had grown up in the era of the voluptuous screen goddesses like Rita Hayworth, Gina Lollobrigida, Jane Russell, and Sophia Loren. We watched the movies so much that, to this day, my sisters and I can do a mean rendition of the debutante dancing with Marlon Brando in the film *A Countess from Hong Kong.* We even quote verbatim, "Daddy always says that dancing stimulates the soul and gives us a desii-iiiiiiiire."

We always said that line while we were dancing, just like the actress who said it to Marlon Brando. When we came to the "desire" part we would shimmy our shoulders back and forth suggestively to our stand-in for Marlon, usually one of us sisters.

> "You might have loved me, if you had known me. If you had ever known my mind. If you had walked through my dreams and memories. Who knows what treasures you might have found. Yes, you might have loved me. If only you had taken the time."
>
> — *Unknown*

When Rosemary Clooney and Vera-Ellen sang the "Sisters" number in the Bing Crosby/Danny Kaye film *White Christmas,* my sisters and I had the blue-feathered fan act down pat, singing our favorite line, "Sisters, sisters, there could never be two closer sisters."

When Halloween came, most of the kids were Cinderella or Superman or a ghost, but my sisters and I were Zsa Zsa Gabor or Marilyn Monroe. I can only imagine what my childhood friend Loretta Olsen thought when I went to her Halloween party as Phyllis Diller, sporting a cigarette holder with a cigarette and yelling, "Oh Fang, where's Fang?"

And, of course, Marilyn Monroe, the most voluptuous of all, with a figure that by today's standards seems flat-out plump, was my mother's all-time favorite. Our favorite Marilyn films were *Bus Stop,* where she played a character who baby-talked, and *Some Like It Hot,* where she was our blond bubble-headed Barbie come to life. We would scream with laughter over the antics of Tony Curtis and Jack Lemmon in those skirts and high heels.

> "If you're going to tell people the truth, be funny or they'll kill you."
> — *Billy Wilder, Austrian-born motion picture scenarist, director, and producer*

That August in 1962 when Marilyn died, my mother was inconsolable in her grief. She cried and cried and cried over poor Marilyn. We cried, too. Poor, poor Marilyn. What was Mother to do without her Marilyn? Then we remembered she still had Jackie Kennedy.

Back when my mother was in California, starlets were getting taller and skinnier and, unfortunately for my five-foot-three mother with her 36-21-36 figure, the studios were losing interest in short, curvy brunettes. Still, she kept trying, until, after she had been in California a couple of years, my father went all the way out there to propose to her and marry her. How romantic is that?

Daddy had fallen in love with her when she was thirteen years old and he was sixteen. It was love at first sight, she said. It was all so dreamy that we would beg her to tell us that story again and again. Today, I probably would think it was sick that a high-school boy was checking out a seventh-grader. But, looking back at photos of my father and mother, it was clear that my mother looked old beyond her years. My father told me that Mother once told him a boy came out to get her for a date. When he got there, her mother, Helen, whom we called Mudder, or Mudd for short, answered the door. She pointed out in back of the barn and told him, "Mary's out in the outhouse — she'll be back shortly."

Mother was horrified that her mother had told her date where she had gone. But my daddy, regardless of her country folks and oil-field shack of a house that we all called Outhome, did not think a thing of it. He had already set his sights on her, regardless of where she lived. Sometimes he told me he would go to pick her up for a date and another guy would already be there. Mother evidently had many beaus.

He never gave up on my mother, no matter how many boyfriends she had. Funny, my mother never told me the same thing about him. Mother always told us that when he finished his tour of duty, he came straight out to Hollywood to find her and propose. With her dream of being discovered fading, she was more than willing to accept the offer.

"Besides," she said, "your daddy was so good looking and so was I. I just knew we would have beautiful children."

She told me how they stayed out West for a while. Daddy had gotten a job working for a construction company that made swimming pools. He enjoyed the work and they had a lot of couple friends. They would all get together and play volleyball or sit around and talk. We girls always asked them if it was so fun out in California, why in the world did they ever move back to ole boring Eureka? Mother always said they moved back because she got pregnant with me and they felt they needed to be closer to home.

I have no doubt that my mother returned to Eureka with her head held high — she was always one for keeping up appearances — but I don't think she ever got over the disappointment of not making it in Hollywood. Unfortunately, my sisters, my father, and I became lightning rods for that disappointment; it thundered down on whoever was in the line of fire.

My daddy had his own dreams. Mother always told his story, too, as my daddy didn't talk to me much. He was always gone, working in the oil fields. She would tell us that he was a high-school athlete. He had won a full football scholarship to Eldorado Junior College and was to be their starting quarterback. Unfortunately, in the very first scrimmage of the very first game of his very first college season, he broke his leg. It was a serious fracture, and when it became clear to his coaches that he would be out for the season, he lost his scholarship and Butler County lost its star athlete. We girls would always say, "Poor Daddy, poor, poor Daddy." No wonder he always seemed so mad.

There were problems in the marriage right from the start; this I can say firsthand. My mother was always a stay-at-home mom. She never worked and considered women who did beneath her in social standing. According to her, the only women who worked were women who couldn't find a man willing to support them. "Always marry rich, girls. It is just as easy to fall in love with a rich man as a poor man."

Mother must not have taken her own words of advice, as it always seemed like we were strapped for cash.

My daddy had channeled his dream of playing college football into one of being a high-school coach. Soon after they settled in Eureka, he enrolled at Emporia State Teachers College so he could get a teaching degree, taking classes during the day and working on the oil rigs at night. Then Mother found out she was pregnant for the third time, with my youngest sister, Karol. Daddy finally had to quit school just one semester shy of his college graduation to go work double shifts to keep up with the increasing demands of maintaining my mother's lifestyle and us kids.

I can only imagine how this weighed on him over the years. Now there was more money. While this should have eased the tension between them, it didn't. Over the years their fights got louder and more frequent. We girls would hide under the covers or sometimes under the bed, or, at best, we'd escape all the fighting by going outside.

For years, I thought that although they fought like cats and dogs, my parents had a common bond in their shared disappointments. Daddy never did become a football coach and Mother never became a famous actress. They were frustrated and, like many married couples, they took their frustrations out on each other and, eventually, on us. My mother always seemed more interested in maintaining her glamorous image than in tending to her three daughters. Oh, she took good care of us when we were babies and really little. But as we got older and developed our own personalities and needs and they did not fit into her neat little box of what we should be like, it seemed like she pretty much just gave up and concentrated on herself. By the time I was in kindergarten, I felt like I was pretty much on my own.

> "Home is a place you grow up wanting to leave, and grow old wanting to get back to."
> — *John Ed Pearce*

When I was writing this book, I realized that even though I had heard the stories over and over about how my parents met and got married, I still wasn't really clear about some stories that my sisters and I had shared as adults. Our versions were different, and they just didn't mesh.

My father was visiting me one Thanksgiving and I decided that I would share what I had written so far with Daddy. I asked him if he would mind if I read him what I had written, since I wanted to make sure everything was correct from his perspective. I was more than a little apprehensive.

Daddy and I sat down and I began to read and read and read most of the chapters on my formative years. I read about my years at home with Mother and Daddy, which was, really, the story of my life. When I came to a place in the book dealing more with my adult life, I stopped and looked up for the first time to see my father's reaction. I was reading from my favorite reading chair, my burgundy wingback in our house's great room. Daddy was perched on the edge of the tapestry sofa in his bathrobe, coffee cup in hand. His first words were, "I wish I could write like that. You wrote your story beautifully, Kathy."

It was probably the highest compliment my daddy had ever given me. I was so taken that I was choking back tears until he started laughing. I stopped, in mid-sniffle, looked up, and asked, "What? What's so funny, Daddy? Why in the world are you laughing?"

As he continued to laugh, a kind of silent snickering with his shoulders moving up and down, I started to panic. Then, much to my amazement, he said, "Because as beautifully as you have written your story, you have it all wrong. None of that is true about me and your mother."

You could have knocked me over with a feather. I was speechless. I, who am never at a loss for words, just sat there dumbfounded. As my mind tried to assimilate the information he was giving me, I finally blurted out, "Daddy, how can you say that? What do you mean? Those are the stories Mother has told me my whole life. What do you mean they are untrue?"

As I wrung my hands and sweat started to bead on my lip from the worry, he began to tell me his version of the story. I quickly pulled my laptop into my lap and began to document the highlights of what he was telling me, typing fast and furiously. Paper and pencil came out next to do a timeline. Everything was finally coming into focus. He told me that, first of all, he hadn't broken his leg playing football. He actually broke it playing basketball. After the doctor had diagnosed a break, he was off the court until he was fully recovered. He explained, "I was just this kid when I went to college and, since I couldn't play, I just dropped out. I didn't really think about the consequences, and then I was drafted."

"But, Daddy, what about your dream of becoming a coach? Mother always told me you wanted to be a coach."

"Kathy, I didn't have a clue back then what I wanted to be. All I knew was that I liked to play sports. So if I couldn't play, then I didn't want to go to college."

I could not believe it. He sounded like he was just like me. He did not really know what he wanted to be, either. I sat there with my mouth hanging half open for most of the time as he was talking. He continued, "When I got back from Korea I went back and enrolled at Emporia State on the GI Bill. Your mother was already in school there and I just wanted to be around her. At the end of the school year, when she went out to California with her sister, Teenie, I eventually just followed her out there."

Mother hadn't gone out to California alone? I knew that she had stayed out there with her half-sister, Glenna, but Aunt Teenie went too? Mother had never mentioned Aunt Teenie going.

I asked my sister about this later and she told me, "Mother went out to Hollywood to be a star and stayed in this horrible boardinghouse. She almost starved to death. But Teenie wasn't with her — she went by herself, and was out there years trying to make it as an actress."

Daddy went on with his version of the story. "Your mother was living with her half-sister, Glenna, and I think Teenie got a job in a bar." A bar! I couldn't believe the story my daddy was telling me, and then I began to question not only this story, but the other one, too.

"Your mother and I never really planned on getting married," he continued. "I never even proposed. We just decided to get married one day. I assure you it was reciprocal how we felt about each other. Of that I am sure, regardless what she may have told you and your sisters."

He told me after a while they just decided to move back home to Kansas.

"But wait, Daddy — I was conceived in Kansas? Mother always told me she got pregnant in California. I even moved to California because I always thought I was really supposed to be a California girl."

He laughed as I continued.

"Mother always told me that when she found out she was pregnant with me, you-all decided to move back home. I can't believe that I moved to California and it was all a lie. If I heard that story once, I heard it a hundred times. Are you telling me she just made these stories up?"

Daddy, still laughing, said, "Well, that's your mother. She always had a vivid imagination and could always tell a good story."

> "The children of warriors in our country learn the grace and caution that comes from a permanent sense of estrangement."
>
> — *Pat Conroy*

I couldn't believe what I was hearing. One by one, I went through all the stories I could think of that Mother had told to me and my sisters through the years — countless times. Daddy would just shake his head and then tell me his version. As I put pencil to paper and did the timeline, I had to admit nothing matched up. My mother had been in Los Angeles for, at the most, two months in the summer of 1955, not two years. I was dumbfounded. What do you do when you find out all the stories of your childhood and youth have been total fiction?

I slumped back in my chair and felt just like the main character of the son in *Big Fish* by Daniel Wallace. In the book, the main character's father is dying and his son goes home to make peace with him. The father was a traveling salesman who told these tall tales, everything from how the son was born to how he was foretold how he, himself, would die. All the son wanted was the truth. All the father wanted was to tell stories. The beauty in the story was that the son finally came to understand that his father's storytelling was just as much a part of him as the truth was to the son. When the son pleaded, "Just tell me the truth, I just want to know the truth," the father told his son, "It's the truth to me, son, it's the truth to me."

I also felt as if the house I had been riding in during the tornado had just dumped me flat down on the ground again and I had awakened with a nasty bump on my head. I was no longer in Oz; I was definitely back in Kansas.

"What about my book, Daddy? I don't want people to think my life is all a big lie."

He looked right at me with his baby blue eyes and said, "Oh just leave it, Kathy. I like the way it sounds better anyway."

As I talked about all of this with my sisters and friends who knew my parents, I found that we all had been told different tales by my mother. Even Heidi, my best friend from childhood, told me a story that I had never heard in my entire life. She told me how my mother said that she had won the 1957 pink Cadillac in a Hollywood screen test. She told me my mother had told her that story many times. I had never heard that story before. What other stories that I didn't know had she told people?

When I asked Heidi how Mother could have won a 1957 Cadillac when she was in California in 1955 and didn't even get the Cadillac until 1961, she just shrugged. "Well, that's your mother," she said. "She always was an incredible actress."

I thought about this for a long, long time. Then one day it dawned on me: Everyone has a different perspective on what the truth really is. Remember, no one ever believed Dorothy either. Just like I used to play and make up stories as a child to make life more fun, I think my mother did this. Not to be deceptive, but because if she told her story enough, then it would become real.

Even though I say I always want the truth, I have to admit I loved those stories. My sisters and I were the ones who always begged, "Tell me again, tell me again how you and Daddy met."

My mother was just trying to give us what she perceived we wanted, a great story. No wonder I love books so much and escape into their stories. I lived with the greatest storyteller of all time: my mother.

I will probably never know the real truth. Once when I was in New York, I saw *The Wizard of Oz* at Madison Square Garden. I was attending a sales conference as a book rep and we had gotten the day off because it was Mother's Day. I practically skipped from the Mayflower Hotel across from Central Park to Madison Square Garden. I could hardly wait for the show to begin. Mickey Rooney was to play the Wizard and Eartha Kitt was to portray the Wicked Witch of the West. Since I was alone, I had gotten a really great seat right up front and center. As the musical began I became lost in the show. Then Dorothy landed in Oz and cried, "Oh Toto, I don't believe we're in Kansas anymore." It was a good thing the theater was dark, because tears began streaming down my face as I remembered those times on the divan with my mother and my sisters. It was Mother's Day, and I missed my own mother and my children. By the time Dorothy began clicking her heels and saying, "There's no place like home, there's no place like home," I was a basket case. Mascara running, nose dripping, I tried to leave the theater as unobtrusively as possible. A man outside held the leashes of the two Totos that were there to greet us as we exited. I stopped to pat the little dogs, sobbing all the while. A woman handed me an iris and said, "Happy Mother's Day!" I just sobbed as I stumbled toward the sidewalk and bawled all the way back to my hotel.

The fact of the matter is, I really don't care if my mother made up the whole thing. She was a movie star to me, as real as Judy Garland, and if she wanted the Technicolor version of her life as compared to the bleak black-and-white, then so be it.

Looking back, I know my childhood had its problems, yet I love both of my parents very much. They have been divorced for many years now, which has created different sets of problems for me. I am learning to deal with it. I was lucky to have known both sets of my grandparents. I also got to know three of my great-grandmothers pretty well. The things I remember most were stories shared when we were together. As much as I think I would love to know the truth, I can honestly say now I have decided that what really matters is that we all love each other.

The following books led me to discover that there really is "no place like home," even if home was not exactly like *Father Knows Best* or *Leave It to Beaver,* the tele-

> You cannot choose your parents or how you are raised. You can choose what you do with your life when you are an adult. Life is about choice.

vision shows of my youth. These are books that helped me to understand my family a little better.

Family by J. **California Cooper** Set in the South before the Civil War, the voice is of slave Clora, who is the granddaughter of a slave and whose mother was a slave who killed herself. Clora tries to kill her children, then succeeds in killing herself. The story unfolds as Clora narrates as a spirit watching from above.

Body of Knowledge by **Carol Dawson** Victoria Grace Ransom weighs over 500 pounds and lives in her family's ancestral home in West Texas, where a silent feud has been battling among her family members for decades. She passes the time being told family stories by the maid. This book has one of the most surprising endings I have ever read.

Mad Girls in Love by **Michael Lee West** A young bride and mother gets in a fight with her husband and hits him in the head with a frozen rack of ribs as he tries to drown her in the sink. Thinking that she has accidentally killed her husband, she flees, only to find out she may have left him for dead but he lives — and with revenge. He divorces her and takes full custody of their little girl. The young mother is devastated and all the while her own mother is in a mental institution and writes letters to Pat Nixon in the hopes of getting her grandchild back for her daughter. And that is only the beginning!

Big Fish by **Daniel Wallace** I love this book because it is about a son who just wants to learn the truth from his father, the storyteller. His father has told fanciful tales throughout his life, and now he is dying. Get a box of tissues — I cried like a baby over the ending.

Comfort Creek by **Joyce McDonald** Mama has run off to become a country singer, Daddy has lost his job, and eleven-year-old Quinn and her sisters are living in a swamp without water and electricity. Need I say more?

The Tender Bar by **J. R. Moehringer** This is the memoir of a Pulitzer Prize–winning journalist who essentially grew up in the neighborhood bar. Living in a rundown house overrun by cousins, with a father he can only hear by listening to him on the radio, J.R. escaped to the Irish Catholic bar looking for a father figure. The bar was his saving grace and also his downfall.

***Miss American Pie* by Margaret Sartor** This book is made up of the actual diaries of Margaret from the time she was in the seventh grade to her graduation from high school. What makes this such a compelling read is that she truly captured my time growing up in the 1970s. Too bad I burned my diaries in the trash barrel after I caught my mother reading them. Luckily, we have Margaret's!

CHAPTER 4
BOOKS SAVED ME

"The man who doesn't read good books
has no advantage over the man who
can't read them."
— *Mark Twain*

At school I applied the same strategy for survival that I used at home. I would become invisible. I hated calling attention to myself as a child. Not only was I painfully shy then, but I also hoped to hide the fact that I wasn't getting the hang of reading. If I did not understand something in our reading lessons, I would just let it go. Better not to understand than to go through the humiliation of calling attention to myself. It is ironic now to think about it, but reading did not come easily to me.

I suffered through read-aloud time in school, terrified that the other kids would laugh at my halting style. Although I cannot ever recall this happening, just knowing that all eyes would be on me when my turn came made even the anticipation unbearable. I could not wait for read-aloud time to be over so I could slink back into my seat and fade back into the woodwork. I always chose a seat in the very last row, preferably behind a boy who was bigger and taller than me, so I could hide behind him. My teachers would always move me, though, as they liked to seat you alphabetically back then.

My rule was never to make eye contact with the teacher. You never had to raise your hand in class if you knew the answer, because when you made direct eye contact, they would call on you every time. During class discussions I sweated bullets and prayed my teacher (pretty much all the teachers were women in those days) would not call on me. Not because I didn't know the answer, but because I would have to speak out loud in front of the class. I would become physically sick just thinking about getting up in front of the class. It worked. I didn't get called on too often.

Whether this was a good or bad thing is hard to say. For just about three whole years, from the first through third grades, my teachers really did not pay much attention to me. I did everything I possibly could to avoid calling attention to myself. I suc-

121

ceeded in my goal because I became an invisible student.

> "The way you overcome shyness is to become so wrapped up in something that you forget to be afraid."
> — *Claudia (Lady Bird) Johnson, First Lady (wife of Lyndon B. Johnson)*

Once there was a flicker of recognition that I existed, and I was caught for a moment in my teacher's eye. In the first grade, my teacher called in my mother for a conference. She had finally noticed me when I had these sores on my legs, which were infected mosquito bites. She had called my mother in to speak to her about my legs, because the school nurse had informed my teacher I had impetigo. I needed a doctor's attention immediately. Mother, dressed to the nines, took me out of school and bad-mouthed me, the teacher, and the school nurse all the way to the doctor's office. She was madder than hell that this woman had reprimanded her for what my teacher and the school nurse perceived as medical neglect.

The teacher, ever alert, was watching me. I was no longer sailing under her radar. A strict disciplinarian with high academic standards, she called in my mother once more that year.

"Kathy is just not picking up reading, Mrs. Murphy," she explained. "May I suggest that you get her some reading workbooks?"

My mother smiled and listened politely, nodding her head every now and again to show her concern. Even back then I knew that behind that perfectly innocent smile lay hell to pay, because I had made her look bad in front of a teacher. Needless to say, after our meeting, my reading did not improve.

> "Shyness has a strange element of narcissism, a belief that how we look, how we perform, is truly important to other people."
> — *Andre DuBois*

I continued to lag behind, growing more miserable and more self-conscious about it every year. That is, until I met Mrs. Boulden.

Mrs. Boulden was my fourth-grade teacher. She was a statuesque woman with a big, warm smile and an expansive personality. She was larger than life in every way.

123

She loved clothes and dressed unlike any teacher I had ever seen. At a time when the accepted fashion among Northside Elementary School teachers was suits, preferably in a subdued, pastel shade of blue or rose, Mrs. Boulden wore brightly colored, flowing scarves, ponchos, and large ethnic jewelry. In winter she even wore a big flowing cape. We kids loved her. She was our Mary Poppins and, just like the British nanny with her umbrella landing smack dab in front of the door of her new charges, Mrs. Boulden entered our fourth-grade classroom stern, firm, and definitely eccentric.

Mrs. Boulden was a bit more than the good citizens of Eureka were used to, but she was left to do her job as she saw fit. She was commanding and wealthy. She was a perfect example of the old saying "Money talks." No one ever questioned her in any form or fashion. I know she went to Europe every summer, and she always brought back unusual tins and packages of food for her students to try: smoked oysters, sardines, butter crackers from France, and, one time, escargots.

She was rumored to have been married more than once, each time to a wealthy man. No one knew for sure, and it was the source of endless speculation among our mothers when they gathered at PTA meetings. Just like Mary Poppins, Mrs. Boulden was a mystery. I didn't fully understand the implications of Mrs. Boulden's marital or economic status at the time. I thought of her as older than Methuselah, but now I think she was younger than I am now, probably in her forties.

Her standards were high and we all rose to the occasion. She was passionate about literature and the arts, and she did her best to pass that enthusiasm on to the children of Eureka, who accepted it in varying degrees. As for me, she changed my life completely.

> "To be great is to be misunderstood."
> — *Ralph Waldo Emerson, American poet, essayist, and lecturer*

People who know me today have a hard time believing it when I tell them how timid and shy I was as a child. I actually shudder to think of the direction my life would have taken had Mrs. Boulden not noticed the very quiet girl at the back of the classroom who never checked a book out of the library. One day she took me aside.

"Kathy," she said, "there's a book I'd like

you to read. It's called *Honestly, Katie John.* The main character reminds me of you. I think you will like her. Please give this book a try, shall we?"

I didn't answer, but I guess my face showed that I seriously doubted her, because she added, "Don't worry about how long it takes to read the book. If you are not finished by library period next week, you can check it out again. But, Kathy, please do give it a try."

Her voice was so full of kindness and concern that it would have been mean of me to protest. Wanting to please her, I took the book home and started reading it that night.

Right off the bat, it was different from all the other books I had tried to read. Katie John was my age and, like me, she wasn't like other girls her age. She loved the outdoors and riding her bike, and she thought it was *ridiculous* the way her best friend had taken to wanting to wear a bra and lipstick. *Yuck!* She didn't much care for boys that way even though her other best friend was a boy. She felt misunderstood and just wanted everything to stay the same. Life was changing too fast. She hated the way all the girls had started acting like the sun, moon, and stars revolved around the boys.

Mrs. Boulden had found the perfect book for the perfect tomboy. Instead of reading for a few minutes and then feeling bored, as I usually did, I kept turning the pages until my mother yelled from downstairs, "Lights out!" I could not believe it was already time to go to bed. As I fell asleep, I kept thinking about that book. The next morning, I woke up a little earlier than usual. Before even getting out of bed I opened my book to where I had left off and finished it.

Mrs. Boulden had been right. I didn't just like Katie John; I *was* Katie John. I felt an enormous sense of relief and happiness. For the first time in my life I wasn't an oddball. I knew there had to be other girls out there who were just like me. I really had thought I was the only one who was self-conscious about the way I looked. Once when I was mowing the yard, a neighbor asked my mother, "What boy is mowing your yard, Mary? I could use some good help." My mother never let me forget that story. She repeated it over and over. I secretly had begun to like boys, one boy named Dana Moser in particular, but my mother (probably sensing that I was growing up too fast) would declare to people, "Kathy doesn't like boys. She is far too busy with her studies to be involved with that."

I learned through Katie that you could be a tomboy and like boys too. You did not have to get yourself all made up and focus all your attention on the boys to be a girl. That book made me begin to see that nothing horrible would happen to you if you spoke up for yourself. Except when I was alone with my sisters or friends, I had spent most of my childhood afraid that I would do something wrong and be punished. I wanted to be just like Katie John, so I began to emulate her. I became brave.

> Be brave. Even if you are afraid, only you will know that. Look outward, instead of inward.

The very next morning in school, I bravely walked up to the front of the class before all my classmates to return the book to Mrs. Boulden. Surprised to see me boldly standing in front of her desk, she raised her eyebrows. "You've already finished the book, Kathy?"

"Yes, Mrs. Boulden. And, please, are there any more books just like that one?"

She was beaming, and so was I as she

walked me over to the bookshelf and pointed out the other Katie John books. I even forgot for those few minutes that I was walking in front of the class to get a book. All I could think was that I could not wait to read the next book. I knew Katie John would lead me to a happier life, and she did.

> "A truly good book teaches me better than to read it. I must soon lay it down, and commence living on its hint. What I began as reading, I must finish by acting."
> — *Henry David Thoreau, American essayist, poet, and philosopher*

A couple of years ago I was reading Oprah Winfrey's magazine, *O*. I love to read her back page and enjoy her inspiring stories. That month she was talking about the first book that turned her on to reading. My eyes could not read fast enough. When she was asked to read a book report she had written in front of the whole class, that was the minute she decided that reading was important. She had been rewarded for reading and for writing such a good report. My heart skipped a beat when I saw the title of the book that had turned her on to reading: *Honestly, Katie John* by Mary Calhoun.

In school, all those decades ago, my favorite time was "book report day." While others were griping, some of us early-infected bookworms were impatient to go to the front of the room and tell everybody about our newest favorite book. I feel the same way just before a Pulpwood Queens meeting of my small Gladewater group. But an even bigger thrill awaits when I can meet my favorite authors and "grill" them — just like my favorite women detectives grill their suspects. I have enough reasons to be a Pulpwood Queen to fill my own book.
— *Lois, of the Pulpwood Queens of Gladewater, Texas*

I share that story with everybody I know. That story does toll the bell for putting the right book into the hand of a young reader. Sometimes children wander into my shop as their parents are out shopping, either next door or close by. I take the time to ask them, "What is your favorite book?" If they don't have one, I'll ask them what they like to do for fun. What is their favorite movie? Nine times out of ten, they will name a movie that was based on a book. You should see the look on their faces when I tell them, "Did you know that was a book first?" They can't wait to read it. I will tell them about the movie I saw as a child called *Mary Poppins* starring Julie Andrews. Until I was given *Mary Poppins* as a book, I too had thought it was only a movie. I loved that book, and later I got a Mary Poppins doll that even came with a travel valise and umbrella. Children love that story. I tell them about how my grandfather Dirt loved the Mary Poppins story, too, so much that he about broke both his legs jumping out of the barn with his mother's black umbrella trying to copy Mary Poppins.

I have learned that when children love a story, they want to hear it over and over again. I always tell them with a wink of my eye, "And guess what? The book is always better."

That child can hardly wait to plop down and get reading. I can see in their eyes that I have lost them. They are already thinking of the story and dreaming of reading the book.

Once I discovered Katie John, I was hooked on books and reading. I read every-

> The more that you read, the more things you will know. The more that you learn, the more places you'll go."
> — *Dr. Seuss,*
> *American writer and cartoonist*

thing I could get my hands on. Our school's library actually consisted of bookshelves that lined the main hallway leading to individual classes; we did not have a real library in its own room. Every chance I got as I entered the hallway to and from recess or lunch, I would walk slowly and peruse the books. I had discovered the wonderful world of reading.

When I was in the sixth grade my sisters and I were baptized at the First Christian Church. We wore white gowns over our white slips and were totally immersed in water as Reverend Leigh dipped our bodies in the small swimming pool–like structure in back of the choir loft.

"Do you believe in the Father, the Son, and the Holy Ghost?"

"Yes, I do," I said as I was lowered into the waters of eternal salvation. I was saved.

As my sisters and I left to change back

into our red velveteen empire-waist Sears dresses, I remember thinking that I was now free of sin. I would read the good book from cover to cover. I was on the road to my ultimate destination: heaven.

Yes, books saved me. Of that I am certain. Just as I was saved by the baptism of the holy waters, I have been saved by the power of the written word. I have been fortunate to have had many great teachers through the years. Jesus was a great teacher, and I can still recall my twelve-year-old self looking up to the stained-glass window above the balcony of our church and seeing Jesus with his flock of lambs. The light would shine through in all those beautiful colors: red, blue, purple, and gold.

To this day, I hold all teachers in high regard. Let's face it, folks — our children sometimes spend more quality hours with them than with us. Because of Mrs. Boulden and others, I understand what an impact teachers can have on our lives. That is why reading is so important; it is the basis of all education. A child may be lacking in so many of the basic needs — food, shelter, family, and love — but if you teach that child to read, you have given him the tools to educate himself. You have given him the tools to find his way and salvation. Education is the key to a better life. Now you know why I am on a mission to get America reading.

> You live like this, sheltered, in a delicate world, and you believe you are living. Then you read a book (Lady Chatterley, for instance), or you take a trip, or you talk with Richard, and you discover that you are not living, that you are hibernating. The symptoms of hibernating are easily detectable: first, restlessness. The second symptom (when hibernating becomes dangerous and might degenerate into death): absence of pleasure. That is all. It appears like an innocuous illness. Monotony, boredom, death. Millions live like this (or die like this) without knowing it. They work in offices. They drive a car. They picnic with their families. They raise children. And then some shock treatment takes place, a person, a book, a song, and it awakens them and saves them from death.
> — *Anaïs Nin, French-born American author of novels and short stories*

If any of you know a teacher who has

made a difference in your life or a person who has changed your life, please tell him or her and thank him or her personally. What a difference Mrs. Boulden made in my life. I never got to thank her personally. Even though she has passed on from this world, she will never be forgotten.

> "We shouldn't teach great books; we should teach a love of reading."
> — *B. F. Skinner*

Once I developed an appetite for books I was insatiable. For the rest of that year, Mrs. Boulden had me reading like crazy. After I read all the Katie John books, she introduced me to the classics like *The Adventures of Tom Sawyer* and *Grimm's Fairy Tales*. I worked my way through the Newbery list and remember loving *Island of the Blue Dolphins, Hitty: Her First Hundred Years,* and *Invincible Louisa: The Story of the Author of "Little Women."* After I had read almost all the books on the Northside Elementary School library shelves, Mrs. Boulden encouraged me to go to the Eureka Carnegie Library, located downtown on Main Street. She explained that I would have to get a library card first. To get there I had to ask my mother to take me.

The Eureka Carnegie Library was an imposing building with big stone steps leading up to massive wooden doors. To my child's eyes it was a fortress, and behind it lay another world, adult and mysterious, and also a little bit scary. As I ascended those huge steps, my stomach filled with butterflies. I was nervous, as I had seen only adults entering this building while we drove by on our way downtown. Inside, the first thing I saw was a long front counter and, behind it, a silver-haired librarian with glasses perched on the end of her nose. My mother and I approached the counter, her heels clicking on the highly polished wooden floor.

"May I help you?" inquired the silver-haired lady.

"Yes, Mrs. Carmickle," answered my mother. "My daughter Kathy wishes to receive a library card." Each word was perfectly pronounced and enunciated. My mother had a way of talking that let people know she was to be treated with respect, and since she was in the library today on my business, I was to be treated with respect, too. I was brimming with pride. The librarian handed my mother a form to fill out: just the usual name, address, and phone number were all that was required. When my mother had finished filling out the application, the librarian looked over the counter and down at me.

"Young lady, I am going to need your signature," she said, pushing the blue ballpoint pen across the counter to me. I glanced at my mother, whose smile told me it was okay, as I took the card and pen from the librarian. I had never felt so grown-up and important in my entire life as when I stood on my tiptoes and reached over to sign my name in cursive, all round and loopy.

Thinking we were finished, I turned toward the stairs that led down to the children's library, but the librarian kept talking. She was all business.

> Read. It will make you a better person.

"You may go now, Kathy. Downstairs is our children's library. You may select two books. When you have made your selections, bring them back to my desk and I will check them out for you. You may check books out for up to two weeks, but you must bring them back by the date stamped on the bottom of the card in the back of the book, or there will be a fine. Do you understand these terms?"

I wasn't sure why she was telling me this. If I didn't have the books back in two weeks, it would most likely be because my mother had refused to drive me. But I nodded yes, and she seemed satisfied.

"Good," she said, smiling for the first time. "Now, you may go and select your books."

I looked expectantly at my mother, who told me she would wait for me in the lobby — this was her way of letting me know I'd better not take too long — and then proceeded down the stairs to the basement. I came upon a large room filled with rows and rows of books, more than I'd ever seen in my life, more than I even knew *existed!* In the center were round wooden tables and chairs with picture books piled in the middle of each table. I immediately sat down and started to look at book after marvelous book, completely losing track of time. Soon my mother appeared in the doorway, and since she had other stops to make — to the Rexall Drugstore, Red Owl Stationery, and Zenishek's Department Store — she gave me a look that told me we had to go.

> "It was from my own early experience that I decided there was no use to which money could be applied so productive of good to boys and girls who have good within them and ability and ambition to develop it as the founding of a public library."
> — *Andrew Carnegie,*
> *Scottish-born American industrialist*
> *and philanthropist*

Oh, this was going to be hard — how would I ever pick just two books to check out? I looked at my mother helplessly, but she wore her "I mean business" look. I knew better than to try her patience. I thought about Katie John and asked myself, What would Katie read? I grabbed two books off the table and hurried to catch up with my mother, who was already walking back up the staircase in her high-heeled pumps.

When we got back to the counter upstairs, the librarian stamped my cards with a great flourish, slipped them into the manila envelopes in the backs of the books, handed them back to me, and smiled in approval.

"Good-bye, dear. We will see you in two weeks."

> "A library is a hospital for the mind."
> — *Unknown*

While my mother shopped downtown, I sat in her big pink Cadillac and read. The problem was that by the time we were home I had read both books cover to cover, and now there would be nothing to do for two weeks but look at the pictures! After that, I got wise and started to check out chapter books so I could enjoy them longer. I read all the Caldecott and Newbery Medal–winning books, Nancy Drew and the Hardy Boys, all the books by Louisa May Alcott. Most of all, I loved the Tarzan books by Edgar Rice Burroughs.

Now here was a character I could relate to, a loner who did not talk much. Tarzan was an excellent swimmer, just like me. There was nowhere more special on earth than our many tree houses, because up there I could really feel what it was like to be Tarzan. I lost half the skin on my hands trying to swing out of trees, sometimes with disastrous results. I wasn't very athletic. I would brush off my knees and climb back up onto the boards I had nailed in for a perch, sit, and read for hours nursing my wounds both inside and out.

Thousands of miles away in the heart of Africa, Tarzan was a long way from where he belonged. Feeling different from the other kids, I was, in my mind, also far away from where I belonged — wherever that place might be. I would look down at our pet at the time, usually our dog Pepper, and pretend he was a leopard stalking me as his prey. Tarzan made the most of the resources around him and wasn't defeated by adversity. If Tarzan could be Lord of the Jungle, I could conquer my fears too.

Years later, when I saw a movie in which Tarzan taught himself how to read with alphabet blocks, I was moved to tears. Like the pages in the Tarzan books, those visual images taught me that reading helps us find our place in the world.

Somewhere in the reading I found that glimmer of hope, something that gave me a reason for living. I learned that if nobody cared for me, then I would care for myself; I learned to love myself. Someone once told me how could another person ever love you if you do not love yourself. I did not love myself; in fact, I didn't even like me.

Many decades have come and gone since I was in the fourth grade. I think about the book character, Katie John, and my teacher,

Mrs. Boulden, to this day. Katie John was the character who led me, step by step, through the wonderful world of reading. Mrs. Boulden was one of those exceptional teachers like Mary Poppins, who, if we are lucky, we get at least once, who opens our eyes to a new way of seeing the world. Mrs. Boulden lifted me up as surely as the tornado lifted up Dorothy and Toto. She deposited me in my own Oz: the amazing, extraordinary world of books.

I've had many other great teachers after Mrs. Boulden, but she was the first, and for that reason alone she holds a special place in my heart. I tell people today that our teachers are the unsung heroes of our childhood. It is hard to believe how many people still believe that silly old adage "Those who can, do; those who can't, teach." That is just plain crazy as far as I am concerned. What could be more important than inspiring a child to read her first book and giving her the gift of confidence?

> "Love is a better teacher than duty."
> — *Albert Einstein*

Mrs. Boulden developed in me a lifelong passion for reading and made me a lifelong learner. She made me long to travel to far-off places. Hearing stories of her travels and experiencing her love of reading was like finding a big old treasure chest. I loved every moment of what I was seeing through her eyes and hearing through her voice, and her guidance helped me discover the wonderful world of reading.

My Favorite Books for Children Inspired by My Teachers

Remember, we are all children and youths until we graduate from high school. No matter how we dress or look, I think children should be able to be children and not made to grow up too soon.

Children's Picture Books

Goodnight Moon by **Margaret Wise Brown** There is no better good-night book than *Goodnight Moon.* So reassuring and calming, to this day this book helps me to have sweet dreams.

I Am a Bunny by **Ole Risom** A wonderful, sweet tale of knowing who you are in this world.

Earl the Squirrel by **Don Freeman** I am crazy about these little fuzzy creatures and Earl is as precious as precious can be.

Doctor Dan, the Bandage Man by

Helen Gaspard Call it nostalgia, but pulling a little red wagon and pretending to fix things is all kids dream of. Besides, the book comes with a real bandage, and kids love bandages.

***Harry the Dirty Dog* by Gene Zion** I must have read this book a kazillion times because both my girls went through a phase where they did not like to bathe. Harry did not like to take a bath either, and got so dirty his family no longer recognized him. A bath and a good scrubbing brought Harry back to his family, who loved him dearly. This book made my girls dive into bubble baths.

***Mr. Rabbit and the Lovely Present* by Charlotte Zolotow** Mr. Rabbit was a favorite book of mine when I was younger, and he was such a little gentleman. I love the books of my youth as the stories were simple yet elegant.

***Puss in Boots* translated from Charles Perrault by Marcia Brown** I dressed up my cats all the time as a child, but a cat that walks upright in boots at will is fascinating to children of all ages.

Children's Chapter Books
***The Tale of Peter Rabbit* by Beatrix Potter** A cautionary tale to remind children to

mind their parents told in a way that never seems preachy but scares them just enough to behave.

The Hundred Dresses by Eleanor Estes Another book from my youth that my daughters loved. Did they love the book or love the way I read the book to them? Maybe it was a combination of both.

The Best-Loved Doll by Rebecca Caudill Kids hardly play with dolls anymore unless they are Barbies or Bratz. I wanted my children to know that dolls were much, much more than just toys to dress and undress. They were to be cherished.

My Side of the Mountain by Jean Craighead George This book about a boy who runs away to live inside a tree out in the woods was first read to me by a teacher after lunch for half an hour every day. We could hardly wait for noon recess to end so we could hear the next chapter of this book.

Charlotte's Web by E. B. White Also read to my class by a teacher. To this day I cry when Charlotte dies and all her babies are borne to the wind. A great book for introducing the circle of life to children.

Island of the Blue Dolphins by Scott O'Dell What would you do if you were left alone on an island for eighteen years? Read this book and find out.

The Secret Garden **by Frances Hodgson Burnett** A lock, a key, a secret garden. Even as an adult I am still under the magic of its spell and want to reread this book.

Books for Teens

Little Women **by Louisa May Alcott** Jo March and her sisters, with beloved Marmee, just makes me long to write and will appeal to those trying to find their place in the world.

Alice's Adventures in Wonderland **by Lewis Carroll** One of my favorite books — I loved going down the rabbit hole with Alice. Kids will love this, too.

Selected Poems and Tales **by Edgar Allan Poe** I have carried the tales and poems of Poe with me through adulthood. To remind me of the poem, I have a stuffed raven who mysteriously appears in different places in my shop.

The Phantom of the Opera **by Gaston Leroux** I have never been so fascinated by anything so scary in my entire life. Teens will love the twisted love story and drama.

Children's books that I have loved sharing with children as an adult!

Children's Picture Books

Madeline **by Ludwig Bemelmans** I am not sure if it is the rhythm of the words or the darling illustrations, but the Madeline books were my and my daughters' absolute favorites.

Lyle, Lyle, Crocodile **by Bernard Waber** Lyle is a crocodile who gets taken from his small charge only to return for a happy ending.

Miss Spider's Tea Party **by David Kirk** Spiders and bugs having a tea party. Both boys and girls love this book.

Chrysanthemum **by Kevin Henkes** What happens when a little girl mouse starts kindergarten and everybody makes fun of her name? Learning to accept what is different is a great lesson.

Art Dog **by Thacher Hurd** A superhero dog who paints, has a cool car, and gets to wear a mask. What more could a kid want?

Weslandia **by Paul Fleischman** A boy who is different creates his own world and his own language, then others discover and accept him.

Olive, the Other Reindeer **by J. Otto Seibold and Vivian Walsh** A play on words that will have the children laughing and loving Olive, the other "reindeer" who saves Santa's day.

***Red Ranger Came Calling* by Berkeley Breathed** One children's book that adults love too. It has a shocking and miraculous ending, and it's great to share with the whole family. It's a must for understanding the true meaning of Christmas. You must believe!

***My Dream of Martin Luther King* by Faith Ringgold** An important book on an important man that conveys what he stood for when up against so many. We should all share Martin Luther King, Jr.'s dream.

***Santa Calls* by William Joyce** This Christmas story will bring siblings from rivalry to best friends.

Children's Chapter Books

***My Louisiana Sky* by Kimberly Willis Holt** A gifted little girl must choose between living with her mentally slow parents and her glamorous and successful aunt in the city. One of the best children's chapter books by a first-time author I have ever read.

***The Witches* by Roald Dahl** The only dolls I truly loved as a child were "Dahl" books. Scary, real, page-turning fun about "real" witches who look like you or me!

***Madlenka* by Peter Sis** The author/ illustrator wrote this book for his daughter who, too, had a fascination for the Madeline books. Here he takes those books one step farther. A visual treat and what a fantastic way to see the world!

Harris and Me **by Gary Paulsen** Post–World War II story about a young boy who was literally farmed out to a relative one summer. Based on Gary's own life, you and your children will die laughing at the escapades of motorizing the bicycle and peeing on the farm's electric fence.

CHAPTER 5
TO KILL A
MOCKINGBIRD

" 'Mockingbirds don't do one thing but . . .
sing their hearts out for us. That's why it's
a sin to kill a mockingbird.' "
— *Miss Maudie from* To Kill a Mockingbird
by Harper Lee

When I was growing up, if a kid had ever made fun of me or told me I was poor, I would have probably knocked him in the dirt and scuffled just like Scout did in *To Kill a Mockingbird.* In our family, we never thought such a thing. We had a roof over our heads. Even if we did go hungry, food came eventually. I learned, though, at an early age that my family was different from most.

We moved a lot, for one thing, from rental house to rental house. One home we lived in when I was just starting school was on St. Nicholas Street. This house was considered to be in the good part of town, on the right side of the tracks. And this was the house across the street from one of the Babson Midwest Institute all-male dormitories.

My daddy bought my mother a used 1957 pink Cadillac and gave it to her when we moved there. I will never forget the day he brought the car home and presented it to my mother. Mother absolutely loved that car. We all did. Good Golly, Miss Molly — the only person I ever knew who had a pink Cadillac was none other than Elvis Presley. We had arrived.

Our old car was a box of heavy metal, a bluebird-blue Ford Fairlane, that had a three speed on the column without power steering. We were always running across the street to get the men from the dormitory (we called them Babos) to push our car down the street so my mother could pop the clutch to get that monstrosity to go in fits and starts. The pink Cadillac, on the other hand, was a dream come true. Everything worked, and it always started on the first turn of the key.

The interior still smelled brand new with white leather seats, accented with glittery silver upholstery, all chrome and as big as a whale on the inside. Mother felt like a queen when she drove that car. Both my daddy and Mother kept it meticulously clean. The first time we went for a drive we discovered how much fun it was just to ride around.

While we were driving, my parents would take turns picking out houses they would really like to own instead of rent. We three girls would ride in the backseat. I was always on the left, Karen was on the right, and Karol would sit on the pull-down armrest in the center of the backseat. I don't think we ever wore seat belts. It only took a couple of times of my mother hitting the brakes to learn how to brace yourself for sudden stops. At times her arm would flail out to keep us from flying through the windshield. I liked to hold on to the little leather strap that hung down from above the door. Fun times were always happening in that car.

As a special treat, if we were really good, we would get to go to the Highway 54 Drive-In, where my aunt Mable worked the ticket booth. Mother would make us take our baths first, then we would put on the baby-doll pajamas my grandmother Murphy had made for us. I can still hear the crunch of the gravel as we would slowly pull into the drive-in and park somewhere around one of the center speakers. We always parked in front of the concession stand. Everyone knew that the back rows were for couples who had come to *park.* As soon as we stopped, we were out the door running as fast as our bare feet would take us on the grass and gravel paths to the playground in front of the massive movie screen. All the kids would play in their pajamas, swinging on the swings or riding on the merry-go-round until the screen would be lit up with the opening Road Runner or Bugs Bunny cartoon. At that signal we would run all the way back to the car.

My sister Karol would lie in the back window to watch, while Karen and I sat in the backseat munching on popcorn, careful not to get one kernel on the floor. Mother would also bring iced tea, but Daddy would splurge and go buy us Cokes and a bag of his favorites — Boston Baked Beans or Orange Slices. We watched movies like *Beach Blanket Bingo* or an Elvis Presley film, and we danced around to *Viva Las Vegas.*

Most of the time the films were double features, and my sisters would fall asleep sometime during the second movie. I always stayed awake. I didn't want to miss one minute of the magic that was playing before me. When we got home, Mother and Daddy always carried in my sisters. I had to walk into the house. Sometimes I would pretend to be asleep when we got there, but they would still wake me. I always wanted to be carried. I guess I was just too big.

When my daddy was home and not traveling to work on the oil rigs, we would go for Sunday drives. Sometimes, if times were good, we would stop at the LoMar Drive-In and get hamburgers, french fries, and Cokes. LoMar was named after the owners, whose names were Lawrence and Mary. We kids always thought that was funny, because those were our parents' names, too.

Years later, when my parents divorced, we girls all wanted to know who got the Cadillac. I think they were supposed to sell it in the divorce proceedings. Since neither one could bear to get rid of that car, they decided on joint custody. Mother had it first. Daddy has it now. He gets it out of his garage now and then to drive in parades. Even though they parted ways, neither one of them could ever part with that car.

When Daddy was away, my mother took to washing the car in the driveway. No big deal, except that she would wash it in her polka-dot bikini. Not too long after the first squirt of the water hose, the Babos would make their way over to our house to introduce themselves to our mother.

Sometimes I would come home from school and a Babo would be there teaching my mother Spanish or the guitar, so I'd just run outside to play until dark. My little sisters were always in tow.

My mother always introduced us to Eduardo or whoever was sitting on the divan with her. Even though they looked different from us and talked funny, we were always taught to treat others as you would want to be treated. I think of Atticus in *To Kill a Mockingbird,* explaining to Scout never to make fun of a guest as the boy from school poured maple syrup all over the prepared lunch at the Finches' house. We always were taught to have good manners and to be respectful of adults. If we were rude or ill-mannered, we'd be clobbered later for making my mother look bad. Mother was good friends too with the president of the University, Dale Baker, and his wife, Helen. I used to play sometimes with their daughter, Carol, who was my age.

Much like Scout, Jeb, and Dill in *To Kill a Mockingbird,* we were too busy running off around the neighborhood. We didn't have a Calpurnia, like Scout's black housekeeper, to keep an eye on us, though. We just pretty much ran wild. With the help of a neighbor boy, John Teter, we had begun to build brick tunnels in the old garage of the dormitory. We spent hours crawling through these brick mazes, scared out of our minds at what each turn would bring.

We had some new girls move in right behind us. The middle daughter, Heidi, was my age, and the youngest, Erica, was my sister Karen's age. They had a fenced-in backyard we would play in. As much as we played over there with the Surber girls, they were never allowed to play in our yard. In fact, no one ever played in our yard. We had to go to other kids' houses because our mother did not like to have children around. I never thought much about it. We just roamed the neighborhood until we heard our mother ring our old school bell, then we'd make our way home way after dark. We would always run like crazy to get home when we heard that bell. I just wish that we'd had an Atticus Finch to come home to when we got there.

And then, suddenly, we moved.

I had just started my second week of the third grade when we came home to find our mother packing. We were moving to a house on Elm Street, right across from the nursing home near the hospital. Going from Mulberry School to Northside Elementary in this new neighborhood was culture shock. Northside students had parents on welfare, single-parent families, and dads who were truckers, grocery-store clerks, or

> "You have to have a dream, you have to have a vision, and you have to set a goal for yourself that might even scare you a little because sometimes that seems far beyond your reach. Then I think you have to develop a kind of resistance to rejection, and to the disappointments that are sure to come your way."
>
> — *Gregory Peck*

gas-station attendants. Since we were on the edge of town, our new school had a large contingency of country kids. Everyone was very friendly and open. I liked the school and the teachers so much better.

Mrs. Peterson, the principal of Northside Elementary, appeared very stern and looked like she meant business. She also was very fair and kind to me. The teachers all looked pretty much the same with their suits and pumps, but their demeanor and manner were entirely different from what I was used to. They really seemed to like us children.

My sisters and I watched over the nursing home across the street very much the same way Scout and her cohorts did over the Radley place. That nursing home gave us the heebie-jeebies, too. We had many conversations about what really went on inside their double glass doors. Though our new house was nice, it was on the wrong side of the tracks, the wrong side of town. I never knew that then. To me, everything was just more relaxed. The kids were friendlier. I made friends pretty easily with one girl in particular named Debbie Teegardin. We became best friends just like Scout and Dill. I practically lived at her house. Our favorite color was blue. We liked exactly the same things: riding our bikes, being outdoors, and being in the Girl Scouts. We were pretty much inseparable. I would call her on the weekends and ask her if I could spend the night. She'd then ask her mother, who always said yes, and then I'd ask my mother, who always said yes. In all the time we lived in that neighborhood, I don't ever remember Debbie — or anyone else — spending the night with me at our house.

I spent almost every weekend at the Teegardins'. The next week we would go to school, until the weekend when I could escape again. When I spent the night at Debbie's, her mother would always cook something delicious: banana nut bread or chocolate–peanut butter cookies. Their family had three meals a day, breakfast, lunch, and dinner, and they were always home-cooked and delicious. Once, when her sister handed me a bowl of peas, my hand slipped and the bowl fell to the floor. I froze. Now I was going to get it. Her mother just came around the table and said, "It's just peas, Kathy. The bowl didn't even break. Finish your meal while I clean this mess up."

I couldn't believe it. I didn't even get a whipping.

Debbie always wanted to be a nurse. She had a big bag of plastic hypodermic kits — without the needles, of course. We took turns being the patient or the nurse. We also spent our time reading Laura Ingalls Wilder books and were always in some sort of construction on a tree house or a clubhouse.

Debbie's daddy was a trucker who hauled cows. One time he didn't see Debbie as he backed up. Debbie tried to get out of the way, but she tripped and fell. He drove right over her arm in that heavy rig. Debbie used to tell me how blood had spurted out between her fingers but that she wasn't hurt badly. In fact, since she was quite young and the ground was soft, she didn't even break a bone. She loved to show me the scars from the stitches between her fingers.

I don't ever remember a harsh word coming from that house. Debbie's mother would sometimes just hug me as she took my torn clothes and repaired them on her sewing machine. When I read *Little Women* by Louisa May Alcott, I always pictured the Marches' home being just like the Teegardins'. Mrs. Teegardin was my Marmee.

My sisters and I never knew what to expect when we came home from school. Most days Mother would still be in bed in her see-through baby-doll pajamas. We loved it when we would come home and find her dressed with her makeup on and her hair done. She was always nicer when she had her makeup on.

In the rare times when our daddy was home, we never knew what to expect from our parents. When my father came home from the oil rigs, it would be all lovey-dovey the first night. My mother and daddy would kiss each other, and Mother would actually cook him dinner. The next day he would either re-tip a drill bit in his workshop for extra money or work on the car. A mentally retarded boy often came by to talk to my daddy when he was working on the cars. Daddy always told us he wouldn't harm a flea, so we were to be polite and not make fun of him. At times Daddy could be just like Atticus Finch.

By nightfall, the whole situation may have changed, and we would crawl under our beds or under our covers, muffling our ears as our parents got into a fight about something. We didn't know what. We didn't understand. We just tried to tune it out.

Under the bed I covered my ears and read with a flashlight. Sometimes we would color coloring books. Mostly, we'd hide until we could fall asleep. One time I woke up in the middle of the night. I was in the backseat of the Cadillac and no one had even missed me. I knew the door was locked, so I just stayed in the car on the floorboards until I heard my daddy leave the next morning.

One time my daddy came home early and started yelling for us to come home in the middle of the day. I knew we were going to get a whipping, so my sisters and I hang-dogged it all the way back home. As we rounded the corner of the house to go into the backyard, we were expecting to see our daddy holding a board for the spanking. What we found was my daddy standing right beside a brand-new turquoise-blue Schwinn bicycle. He had bought us a bike for no reason. None of us had a birthday, it wasn't Easter or Christmas — just an ordinary day. We all had to share, but who cared? We had a bike! We were mobile.

One year for Christmas I had been hoping to get a Victorian doll from the Sears catalog. I got a globe and a crinoline slip from Santa. I don't remember what Karen got, but she wanted a slip and didn't get one. Karol got a Barbie case but no Barbie. I now collect globes, symbolic of how far I have come, or maybe this just shows my perverse and distorted sense of humor. Don't own a single slip, though. They're still too itchy.

Then when I was in the fifth grade, I got a Smith Corona typewriter from Santa Claus. This will always remind me of the wonderful book *Christmas Past* by Bill Duncliffe. It wasn't a play typewriter, but the real thing. I still have it and I taught myself to type on that thing by studying the book that came with it. For years, I typed and pretended I was going to become a writer just like Laura Ingalls Wilder. Later on I was Harper Lee or Truman Capote, depending upon my frame of mind that day as I typed up in my tree house. I had a hard go of it because it seemed that nothing really cool ever happened to me when I was a kid. My Smith Corona still has the original typewriter ribbon on it, and if keys could talk, oh honey! Now I collect old typewriters and typewriter keys too! I have one in my shop that holds paper in the roll announcing the next author event.

> "There is nothing to writing. All you do is sit down at a typewriter and bleed."
> — *Ernest Hemingway*

Nursing-home patients often came to our house asking to use the phone. They would tell my mother that the nursing-home people were trying to kill them and they had to call their daughter or son. Mother always let them in to use the phone and would pour them a cup of coffee. Pretty soon orderlies dressed in white would arrive to haul them back crying, kicking, and screaming. One old couple who always escaped liked to sit in the ditch in front of the nursing home holding hands. Like Scout, Dill, and Jem, we were forever watching the world unfolding before us from our hiding spot in the bushes, or up in the tree house, where we were always making improvements by adding shelves for books or driving nails into the tree for a place to hang our jackets.

When I was just about to start the sixth grade we moved again, and this time into a house that had once been a barn. This was ultramodern and very hip for 1968. We loved this new barn of a house, with its sliding-glass patio doors and real working limestone fireplace. We had now moved back to the old neighborhood, the neighborhood with the Mulberry School that was now Eureka Junior High School. I couldn't wait to go back to school and rejoin all my old friends.

But I soon found out that I wasn't going to be able to pick up where I had left off.

As I began the sixth grade, I was informed that I was behind. The caliber of teaching at Northside Elementary School wasn't up to the standards of Eureka Junior High. I had always been a good student, getting mostly As and some Bs, so I picked up the pace. I would study even more.

> "Until I feared I would lose it, I never loved to read. One does not love breathing."
> — Harper Lee

I saw my friends from St. Nicholas and they hardly spoke to me. "Oh, hi," they might casually mutter as they ran off to join their friends on the playground. I was a little pudgier since they had last seen me as a rail-thin third-grader. I began to take solace in food. To hide my increasing size, I would wear a black wide-waled corduroy suit that my Grandmother Murphy had made for me. I wore it with a different blouse just about every day. School had just started and it was August in Kansas. The weather was hot, but there was no way I was going to take off my jacket and wear just my paisley blouse. I would just sweat in misery. Mulberry School was not air-conditioned. I was miserable. I spent most of my time studying my books and reading. What friends I had were in books, and I stuck with those tried-and-true friends.

When Scout went to school for the first time and had to wear a dress, she didn't take to the rules and regulations very well. She ended up in a scuffle with a boy and was severely reprimanded by her father. While I never quite had that happen, I can assure you I was just as miserable as Scout was. Both Scout and I were the piece of the puzzle that just didn't fit.

Even though the town I lived in was all white, I learned right away that there were other forms of prejudice besides color. We had moved to the wrong side of the tracks and that was a major no-no. Living in a beautiful new home on the right side of the tracks now didn't seem to make a lick of difference. My parents had joined the country club. We drove a beautiful car. What was the deal? One former friend even blurted out, "Your mother is wild. She hangs out with the Babos and we heard she sunbathes in the nude. Ewww!"

"She does not! She just unhooks the strap so she won't get tan lines, and the Babos are our friends," I retorted as they all ran off laughing and skipping to the playground. I knew exactly what Dill must have felt like when asked who his father really was. The others just didn't understand. My mother hung out with the Babos because she was learning Spanish, and the guitar. I told myself that these kids just did not understand the idea of bettering yourself, but secretly I was ashamed of the way my mother carried on.

It didn't help when my world-history teacher announced that the capital of Colombia was Bow-gut-a. I immediately raised my hand and told him, "It's pronounced Bo-ga-toe."

The kids in the class all started whispering. Who did I think I was, correcting the teacher? I was excommunicated.

There came a day I will never forget. I did not have any underwear. I had begged my mother to buy me some, but she told me that we could not afford anything new. We had just bought a house. I was forced to wear my baby-doll pajama bottoms to school, washing them out each night so they would be clean. Our gym teacher made us take a shower each day after class. (I swear, gym has probably done more to destroy children's egos through the years than to create a basis for lifetime fitness.) I was hurrying to the shower with my towel wrapped tightly around me when a really popular girl spied a bit of the ruffle showing on the side my towel didn't quite cover.

"What have you got on, Murphy?" she asked as she whipped back the towel. Laughing and bending over, she yelped, "Oh my God, she has on her pajama bottoms. What a baby! I bet you don't even wear a bra!" She grabbed the towel and lifted it higher to show that she was right. Humiliated, I hid my tears in the spray of the shower. I waited until everybody had left before I got out, even though I knew I would be late for class. I was mortified.

Scout was just like me. What was all this big to-do over bras and growing up? I was now in the eighth grade and I just wanted things to stay the same, to stay a kid. I had seen what happened when you became a grown-up. You fought all the time. I would rather have had everything just stay the same and be a kid.

The year I was supposed to start ninth grade, I decided I had had enough. I do not know if it was puberty kicking in or something else. I just decided that if I couldn't beat them, I'd show them. I had my mother give me a makeover.

She began by plucking my unibrow into two perfectly arched brows. Already I looked older. She taught me about good skin care and selecting the perfect foundation — on this she was an expert. She even bought me my own makeup kit from the Rexall Drugstore, and not the cheap dime-store kind, but Max Factor — real, grown-up makeup.

I had shot up to over five-foot-eight the summer before and had started using Sun-in on my hair at the country-club pool, which brightened my dishwater-blond hair back up to a sunny blond. I had a nice tan from endless hours baking in the sun with my Johnson's Baby Oil saturated with drops of iodine. I was allergic to the ever-popular Coppertone. It gave me a rash. I had good skin and a great tan, so I raided my mother's closet.

I found one of her old white twirling skirts that came right to her knee when she was in high school. After I rolled the waist band three times, the skirt gave me a perfect twirly miniskirt. I chose a lavender Tom Jones blouse with a wide collar and wide French cuffs. I borrowed a pair of my daddy's cuff links with the monogram M for Murphy. I sneaked a pair of my mother's pantyhose from her drawer, then locked myself in the bathroom and shaved my legs and underarms. With my purple square-toed and silver-buckled end-of-the-summer-sale shoes, I had an outfit that would have made Marcia Brady cringe with envy.

The first day of ninth grade, I tied my hair with a thick purple-yarn bow, applied my lavender eye shadow and frosted pink Flame Glo lip gloss, got dressed, and looked in the mirror. Thinking I looked pretty darn good, I marched out the door and immediately went to the house of my childhood friend Heidi Surber.

After I knocked, Mrs. Surber opened the door and said, "Why, Kathy, haven't you grown up this summer? Heidi, Kathy Murphy is here."

Heidi came downstairs and said, "Hi, Kathy, ready to go to school?"

That day was the best day of my life. First, I was happy because Heidi didn't say a thing about my miraculous transformation from a geeky, chubby eighth-grader. Second, I knew I would be accepted into junior high at last.

It is amazing what a little makeup, a new do, and a great outfit can accomplish. I learned something that day: No matter what dire circumstances life throws at you, put your best foot forward. Have a positive attitude, accentuate the positive, be the kind of person you would want to know, and leave the negative behind. You have to believe in yourself. I believe, oh, I believe.

My parents, though wrong about a lot of things, had it very right when it came to some of the most important ones. My mother taught me that it is okay to be different, to have an artistic flair. She taught us all to have good manners, and to be kind to older people. Never once did she turn away from a person who came to her door in need. She taught us how to act in public. She gave all my sisters a love of the arts, and gave me a deep desire for literature, music, theater, and travel.

My daddy taught me that you treat everyone with the respect they deserve, regardless of race, color, religion, whether or not they are mentally or physically handicapped. I still often think of my daddy, a fresh Kansas kid barely out of high school, being sent to Korea. He has told me story after story about what it was like over there, for the Koreans and for the Japanese. He also told me he got picked on by some boys from back East, but these good old Southern boys took him under their wing and protected him. These good old Southern boys were black, and he trusted them with his life.

"Never judge a man by the color of his skin," he used to tell me, "only by the content of his character." I have not forgotten his lessons, or my mother's. Those are the stories I choose to hold close to my heart.

> Never judge people by their race, religion, gender, color, economic background, or whether they have had cosmetic enhancements or color-treated hair, but only by the content of their character.

I read *To Kill a Mockingbird* every year, sometimes twice. I read it again and again because that book gives me hope and has changed my life for the better. I learned that there is no clear black-and-white line between good and evil. I learned of social injustices and the importance of a moral education. I now know that what happens to us as children does shape us as adults. I also know that we have a choice to let our childhood secrets beat us down or to let those lessons lift us up to a higher calling. I learned that there is no shame in growing up in a small town, and that you never have to apologize for where you come from or the education you have received.

> "The book to read is not the one which thinks for you, but the one which makes you think. No book in the world equals the Bible for that."
> — *Harper Lee, author of*
> To Kill a Mockingbird

Books are as important to me to this day as the sustenance of food for my body. The following books are ones that have fed my body, mind, and soul. They helped me find out who I really am. I have also noticed that they all have a central theme of social injustice. It is time we open our hearts, minds, and souls. Walk in another man's shoes and somehow you will find out we are often not so much different from, as ignorant of, our fellow man.

***The Color Purple* by Alice Walker** I remember reading this book because I was furious Steven Spielberg did not win the Academy Award for best picture. I understand why Oprah wanted to star in the film and why she has personally backed the Broadway play.

***Sugar Cage* by Connie May Fowler** This was Connie's first book and I read it after I discovered *Before Women Had Wings.* I was shocked that a white author could write so well about the black experience. I love books that let me walk in someone else's shoes.

***Jubal* by Gary Penley** I know this is a controversial book, but I saw it as a moving story about people who overcame great diversity. It also taught me that you never know who you are until you know where you came from.

***Daughter of Fortune* by Isabel Allende** Until I read this book, I had never read about anyone from Chile. A chilling tale that will make you understand the import of "Give me your tired, your poor, your huddled masses yearning to breathe free," as inscribed on the Statue of Liberty.

***One Mississippi* by Mark Childress** A quirky book about a teenage boy from the Midwest who is suddenly thrown into high school in the South. Mark does the Southern novel funnier, stranger, and more tragic than anyone else I read today. In his wild romp of a read, he shocks you and makes you think.

***The Sunday Wife* by Cassandra King** From the time my parents bought their house on Plum Street, I grew up next door to my church's parsonage. I learned up close and personal how hard it is to be a minister's family. This story is about a minister's wife who struggles to be heard. Cassandra King gives her a voice.

CHAPTER 6
GRAND OPENING OF BEAUTY AND THE BOOK

"I write for no other purpose than to add to the beauty that now belongs to me. I write a book for no other reason than to add three or four hundred acres to my magnificent estate."
— *Jack London*

My belief in service served me well in my first beauty shop, Town & Country Headquarters. How I got into the hair business is not too complicated, really. As I mentioned before, after I graduated from high school I enrolled at Kansas State University in Manhattan, Kansas, home of the K-State Wildcats. In retrospect, I think that decision was wise, though at the time that was not my original decision. That choice had been made for me.

When I was in school every high-school senior had an appointment to sit down with the guidance counselor sometime before he or she graduated. My meeting with my counselor, Mr. Straight, was not productive.

"Kathy, tell me. What would you like to do upon graduation?" Mr. Straight asked.

"I've been thinking about it quite a bit, sir," I blurted out nervously, "and I think before I continue my education, I would like to travel. I was thinking that maybe I would like to be an airline stewardess. That way I could get to see some of the world, and I could have a job to support myself at the same time while I tried to figure out my future course of study."

"Kathy, which branch of the military would you like to enlist in?"

Confused by this sudden change of venue, I corrected, "Sir, I do not want to join any branch of the military service. I just want to become an airline stewardess. Don't they have special schools for that type of career?"

"The only airline-stewardess training that I know is through one of the wonderful branches of the armed services."

"But, Mr. Straight, one of my parents' best friends' girlfriend is an airline stewardess and she went to this stewardess school — I think in Kansas City."

"I think, young lady, you would get more of your parents' approval if you would attend Kansas State University," he said as he

176

slid the paperwork across the table in front of me. "You are already preapproved to attend this fine university, and I have taken the liberty of filling out the paperwork for you. You were accepted based on your excellent grade-point average. I also chose elementary education as your major."

"But Mr. Straight, I really —"

"No need to thank me, Miss Murphy," he said as he stood up and handed me the paperwork. "I am positive your parents will be pleased to have you attend such a fine institution."

He smiled as he walked around the desk and led me out the door. Since I was seventeen and had not yet found my voice, I took the paperwork and enrolled at Kansas State University in Manhattan, Kansas.

I had always been a good student, though I found college to be a big change. Suddenly, instead of being in classrooms of fifteen to twenty students, I was in auditoriums that held five hundred. I was scared. There were professors who seemed to look down their noses at me, who treated me like I was stupid. I knew I wasn't — I had gotten all As and Bs in high school! — but it sure did throw me to be treated like a country bumpkin.

All the painful shyness and intimidation I

had felt in elementary school returned. I felt just like I had as a girl from the wrong side of the tracks: not good enough. Outwardly, I think all my friends thought I was fine, but in reality I was in over my head. I was drowning in fear and too afraid to tell anybody. I spent my first two years in college feeling tongue-tied and humiliated, afraid of just about everyone and everything.

It didn't help that, when I attended my first English 101 class — always one of my best subjects — I was called up to the instructor after class.

"Miss Murphy?" the professor announced to the classroom. "Would you please see me after class?"

I had raised my hand and nodded yes. After class, I approached her desk. I was wearing cut-off jean short shorts and a halter top, and my long blond hair hung down to my waist. I felt that I was dressed like most of the other students that August in 1974 and looked pretty good with my summer tan from lifeguarding.

She looked up. "Miss Murphy?"

"Yes, that's me."

I stood at her desk with my binder clutched to my chest and my knees starting to shake.

"I don't like your looks. I don't like your

attitude. And if you don't drop this class immediately, I will give you an F. Do you understand what I am saying?"

I looked at her in disbelief. Wasn't I always the teacher's favorite? What had I done wrong? I think I nodded, and then she brushed me out of the classroom.

I walked to my next class in sheer panic. What had I done? I could hardly move, and it felt like everybody knew what had just happened. I hadn't said a word in class, just entered the classroom and sat down toward the back. I quietly laid out my books and got my papers ready. I thought, There must be something wrong with me. I became paranoid that semester.

Oh, I went back to class like a fool when I should have gone to the dean's office and voiced a complaint. I sat in the back thinking I must have imagined the whole thing. The professor must have been kidding. Then when I got my first essay back with a giant red F written at the top, I quickly hid the paper and thought, What am I going to do? There was nothing wrong with my essay. I had followed the teacher's instructions to a T. But I began to doubt myself.

I kept going to class thinking I would try harder on my next paper, but the next one had a big red F on it, too. This continued until I realized that she really was going to flunk me. I tried to drop the class then, but it was too late. Finally I just stopped going. There was no sense in going to a class where the woman obviously hated me for some reason I couldn't figure out.

Looking back, I think I know why she did not like me. She thought I was a snob, when in reality I was scared beyond reason. In Tom Wolfe's *I Am Charlotte Simmons,* young Charlotte too was misunderstood her freshman year, and when I read that book, it was very painful for me. Tom Wolfe got it exactly right for a young college-freshman small-town girl. Mr. Wolfe got some pretty harsh criticism for his book, but coming from that very quiet place, I can tell you, I could have been that girl. I had no one to talk to about my situation. I was so ashamed. I began to hide in my room.

> "If you think education is expensive, try ignorance."
> — *Derek Bok, President,*
> *Harvard University*

Sometimes I would go out with my friends on the weekend. Most of the time I just hid between classes. What is embarrassing for me to admit, even to this day, is that as I went to other classes, if somebody looked at me a little differently or made a comment that was not very nice, I would stop going. Like I said, I became paranoid.

The classes that I loved were art design, theater, music, ballet classes — what most students called the "pud" courses, or courses easy to ace. I excelled in anything having to do with the arts. Those instructors seemed to embrace me. I had one design instructor who was really the only teacher who made eye contact with me. She actually was interested in me as a student and told me I had an innate sense of style. God bless her — that was about the only personal word I received while I was there. Granted, as freshmen you are herded like cattle through every endless and incredibly boring prerequisite course. I was accustomed to teachers who were passionate about their subjects like Mr. Hayward, Mr. Lassiter, and Madame Basham from high school. These teachers lived and breathed their subjects.

The core courses like English, psychology, and biology became almost unbearable for me to attend. First, they were huge, some with over five hundred students in the auditorium. Tests were done on Scantron cards, and I was out of my element and comfort zone. I made all As in the fine arts courses and Ds or Fs in the core-curriculum classes. All I saw when I received that little typed sheet of grades was failure.

That first semester, I made either As or Fs. I was eventually put on academic probation and had to go before a board to explain what was happening. I am sure that they thought I was just a typical student partying too much and not doing the work. That wasn't it at all. When I went before the board all I could do was cry. I don't think I said a word on my behalf as I was reprimanded and told I was wasting my parents' money. I was horrified and embarrassed beyond all realms of my conscious being.

I realize now that I should have gone first to a smaller college. The many times I have gone back to college to take classes later in life I have done extremely well, making straight As and even the dean's honor roll.

> Start out small, do it well, then you will be able to move on to the bigger stuff.

When my mother announced at the end of my sophomore year that my parents would no longer be able to pay my way through school, at first I was relieved. My younger sister Karen was about to start college and my mother informed me that two in college was beyond their budget. Then I suddenly realized that if I did not go to college, what in the world was I going to do with my life? The panic was back.

My mother always told us, "Girls, you will go to college." The details beyond that statement were never made real clear. It never dawned on me that my parents would not be able to pay for us to go to college. I realize now what an incredible amount of money it would take to have three children in college at the same time. My sister Karol would be following right after Karen.

"You will have to get a job, Kathy, and work your way through college."

I had a hard enough time dealing with school alone, but a job too? I was now scared and sick. (Later in my life, my father told me that this was completely untrue. My parents had plenty of money at that time to send all three of us to college. My father was by then a drilling superintendent for a large drilling company, and even with us all three in school, he could well afford to send us all through college. In fact, he even told me that if I wanted to go back to college today, he would pay for it. I thanked him, but declined.)

> Sometimes you learn the most in life when something goes very, very wrong. Remember that no matter how low you go, the only way out is back up.

I was lost. I kept away from my friends because I did not want them to know what was happening. Finally I decided I had better get a profession that would help me finish college. Working at a local hamburger joint was not going to be an option. Even though I had had a horrible experience at college, there was never any other course of action in my mind than eventually to fulfill a college degree.

I racked my brain for an answer to my problem. I remembered that my roommate had had a girlfriend who was going to beauty school. I had gone once to get a haircut and saw that she really seemed to love doing hair. The beauty school was a state-of-the-art Redken school and I was impressed. I wondered if I could get into beauty school.

Luckily for me, God was watching out for me and planted a seed of hope in my brain.

Remembering my daddy's advice about learning a trade with your hands, I headed to downtown Manhattan, Kansas, to inquire about enrolling in Crum's Beauty College, a Redken-sponsored school that taught the latest techniques from Los Angeles to New York. Now all I had to do was convince my mother to send me there.

At first my mother hated the idea. "No. Absolutely not," she told me. "No daughter of *mine* is going to be a beauty operator. Over my dead body."

> We are all given a gift. Find your passion and talent and make it your life's work. You will never work harder in your life, but it will not seem like work. Your work will seem like play.

As far as she was concerned, doing other people's hair was a service profession about one-half step removed from the world's oldest.

"Beauty operators are not the 'best' kind of people, Kathy," she said, meaning that they are loose, trashy women who do nothing but gossip. She hated hairdressers and she was forever reminding me about the one time she went into a local salon to change

185

her black Jackie Kennedy flip into a Marilyn Monroe platinum do instead of doing her own hair as usual. They fried her hair and then laughed about it behind her back, according to my mother. "They did it on purpose as they were just jealous of me." As a child I was scared to death of hairdressers after that and it wasn't until years later when I went to Crum's Beauty College that I ever really got a professional cut and style.

Knowing what I know now, it is no wonder that my mother's hair became overprocessed, since she had been dying her hair jet black with home hair color for years. I am surprised her hair didn't fall out in clumps.

When you go to the salon, tell the stylist exactly what you want and bring photos of the style. The most important ingredient in making someone happy with her hair is good communication. Be specific. Don't expect the stylist to know what you mean by "just a trim."

I waged an all-out campaign to persuade her, explaining that Crum's Beauty College was a state-of-the-art school, very prestigious, and graduates were not considered "beauty operators" but licensed, professional cosmetologists.

I asked my mother if she and Daddy could afford to put me through beauty school since it was only for nine months, not four years.

"No, Kathy. I'm afraid we can't help you."

"I know — we can use the money that is in my savings account," I told my mother.

"Gone. All gone. I have already used that for you to go to school."

"All of it?"

"Yes, all of it is gone."

Every babysitting dollar, every paycheck from working at the country club, any birthday or Christmas money — all went into that savings account. I'd worked for four years every summer and more, and it was all gone.

I quickly told my mother, "If I can earn the money for school, could you at least help me with room and board?"

She said she would think about it.

Now I don't remember how much it cost to go to beauty college, but I worked three jobs that summer to get the money together. I was a state park rangerette, an elementary playground supervisor, and a waitress and cook for my aunt Teenie at the Ember's bowling alley, snack bar, restaurant, and private club.

One night when I was cooking for the snack bar, restaurant, and private club all by myself, my aunt came in and told me we were also cooking steaks for a bank dinner in the private dining room.

I could cook a couple steaks at a time, hamburgers, and chicken-fried steak dinners, but for steaks all cooked differently with all the sides for over thirty diners? My aunt threw on an apron and, by gosh, between the two of us we cooked those thirty steaks served rare to well done and all the other orders for the snack bar, restaurant, and club at the same time.

The bank president came back to the kitchen later and personally gave me a huge tip. I tried to give it to my aunt Teenie but she made me keep it. She was very generous that way.

I finally wore my mother down to the point where she agreed to visit the beauty school and meet Mrs. Crum, which was the best thing that could have happened.

Mrs. Lucille Crum, an imposing woman in her own right, could go head-to-head with my mother in the self-importance department any day. During my mother's visit, Mrs. Crum poured on the haughty charm. My mother ate it up with a spoon and decided that, well, yes, she would consider letting me attend this distinguished beauty school, and wasn't Mrs. Crum just lovely? She finally consented and realized she would not be consigning her eldest daughter to a house of ill repute or even to one of hopeless tackiness. I started at Crum's Beauty College in August 1976.

> "I think the most important thing a woman can have — next to talent, of course — is her hairdresser."
> — *Joan Crawford*

Mrs. Crum was a distant woman. She was a brilliant businesswoman who did not stand for anything but the best. I liked her no-nonsense kind of leadership. Everything was very black-and-white for Mrs. Crum.

> Always have someone outside of your own family to turn to in times of trouble. Become a mentor. You never know the difference you can make in a child's life by steering him in the right direction, giving him your knowledge and experience in your chosen field.

I loved beauty school. I was in my element, since I approached hair as art. Just like the song, "I was gaining confidence daily." I threw myself into my new schooling.

Studying in a mostly all-girls school — with the exception of my friend David — I was having the time of my life, doing something I enjoyed. The work was easy and what we learned in class time was directly related to the job we would be doing. I aced all the tests and, cutting hair, I created styles unique to each individual.

Each student model or client, I discovered, was a blank canvas and you could create a unique look just for her. I had found my life's work in this profession and threw myself entirely into learning all I could to become a licensed cosmetologist. About halfway through the program, we had a student competition. I partnered with another student friend, Jeannie. At the competition we each had to set, then style each other's hair. We both won places in the competition. I won both for the styling and for the modeling. For sure, I was bitten by the hairdressing bug.

Mrs. Crum started to take special notice of me and took me aside to have me do things that the other students weren't allowed to do. She was a commanding presence and she demanded perfection. I was ready to do her bidding. Toward the end of school, we had another competition, and I won first place for my styling and as a model. Mrs. Crum decided to send Jeannie and me to a state competition. We would be competing with professionals in the business. I was never scared, just excited. Mrs. Crum made both Jeannie and me feel like we were up to the task. We had so much fun at the competition, even though we did not place, and Mrs. Crum was very pleased with our performance. We had represented Crum's Beauty College well.

At the end of the nine-month program, I passed my state boards with flying colors — straight As. Mrs. Crum offered me a stylist's job at her salon, Crimper's, near the Kansas State University campus. I declined. I wanted to do something unique, not just be one of her haircutters. A door of opportunity opened even before I got a chance to knock. I got a call from my friend Becky. She had just bought this huge salon, the Hair Shack, in Lawrence, Kansas. Would I like to come and work with her? I was packing my bags the instant I hung up the phone.

In Lawrence I worked with some of the best stylists I have ever known. I stayed there, honing my skills for the next three years, and then I moved back to Eureka to open my first salon, Town & Country Headquarters.

I had a good downtown location in the former Princess Barber Shop, located right next to the Princess Theater, and plenty of potential customers in the men, women, and children on their way to see a movie. With Smokey Robinson singing "Cruisin' " and Kool & the Gang singing "Ladies' Night" on the stereo, I held my grand opening in 1980.

Early on, I received a little heavenly boost. My very first customer was my minister, the Honorable Reverend Gilbert Daniels. I give each and every one of my customers my all, and Reverend Daniels was no different. I took my time with him and we had a nice chat. When he left, I felt great about that first haircut but didn't think too much more about it.

Then, the very next Sunday, the good minister announced in front of God and the entire congregation of the First Christian Church that he had paid a visit to church member Kathy Murphy at her new salon and received the best haircut he had ever had. I was so shocked by his kind words, and what a blessing that statement turned out to be for me! As far as free advertising goes, word of mouth is always the best, and, in a small town, it doesn't get much better than coming from a minister. From there on in, I had all the work I could handle.

> "If truth is beauty, how come nobody gets their hair done at the library?"
> — *Lily Tomlin*

Years later, when I opened Beauty and the Book, I wasn't to receive the official blessing of the church. My family and I hadn't yet joined the First United Methodist Church, of which we are now members. What's more, this time my hair salon/bookstore wasn't even located in the heart of downtown, but in the piney woods on the outskirts of Jefferson, Texas, miles off the beaten path. I worried that customers wouldn't be able to find us.

I had put up a big Beauty and the Book sign in front of our house, out by the blacktopped county road. My logo was elegantly designed by my friend and sign designer Carol Hodges and featured a reclining woman reading a book, inspired by Winslow Homer's watercolor *The New Novel.* I had a smaller version on the glass window in the front door of my new shop.

But a lot of good this was going to do me unless people knew about my shop in the first place.

Most people can go a week or two on vacation and not miss the paycheck, but my salon equipment was late in arriving, and now a month, then two, went by with expenses going out and nothing coming in. The money invested in the shop was gone. I don't mind saying that I had my share of dark moments.

I lay awake all night thinking about past adventures, about risks I had taken. I had started down several yellow brick roads in my time and, like Dorothy, the Tin Man, the Cowardly Lion, and the Scarecrow, had sometimes only narrowly escaped disaster. It was all right when it was just me, but now there was my husband, Jay, and my precious girls. I couldn't sleep for worrying that a big, black twister was going to sweep through and demolish my new venture. Had I finally bitten off more than I could chew? Was I leading my family down a path of no return to financial ruin?

Beside me, Jay, every bit as worried as I was, feigned sleep. I knew he was awake, and he knew that I knew, and I knew that he knew that I knew . . . but we just lay there in silence, conscious of not talking, nursing our own private fears. It was easier to say nothing than to start a conversation that would only wind up an argument.

And then two things happened that gave us the big break.

First, years earlier, when I was still working as a sales rep for Henry Holt Publishers, I had recognized an amazing and talented author, Kimberly Willis Holt. She had written a children's chapter book set not too far from Jefferson, near Alexandria, Louisiana, titled *My Louisiana Sky*. I was a huge fan of the book and I decided to give her a call to introduce myself. I asked her, first, if she was related to Henry Holt of Henry Holt Publishers, and she laughed as she assured me she wasn't. Then I asked her if Henry Holt was touring her for this book. She told me no, she was a first-time author. I asked her if she would like to go to some bookstores where I thought her book and talent would be appreciated. She told me she would love to go to some bookstores. On her own dime and mine, I lined up a little book tour.

We went first to a favorite bookstore of mine, the Book Merchant in Natchitoches, Louisiana, and held her first-ever book signing. We have been best friends ever since. Kimberly has gone on to win award after award for her children's books, and she even won the National Book Award for Children's Literature in New York, with the award presented by none other than the bookselling queen of all queens herself, Oprah Winfrey. Now that I was opening my shop, I asked her if she would come and be the first author to grace my doors. She graciously consented.

> Be a lifelong learner. Every day is a lesson and an opportunity to learn. Take a class, continue your education, and your life will have more meaning and purpose.

At first, I took her around to East Texas schools that have literacy-promoting librarians like my good friend Helen Thompson of Chapel Hill ISD in Mt. Pleasant, Texas, and, folks, we sold some books. Next we had a big dessert reception for her at my brand-new shop. I invited in my Girl Scout troop to help host the event. Kimberly was the first author to let me give her a full makeover, and we did it in the first part of the evening's event. We sold enough books that my store had a beginning that was certainly in the black. I will be forever indebted to this faithful friend who has held my hand numerous times on the precarious ledge we call bookselling.

> "A true friend is someone who thinks that you are a good egg even though he knows you are slightly cracked."
> — *Bernard Meltzer*

The second piece of good fortune also happened around the time of our grand opening. I had sent a press release to *Oxford American,* the Magazine of Good Southern Writing, to announce the opening of my shop. I actually sent the press release to everybody I could think of, from the local weekly newspaper, the *Jefferson Jimplecute,* to my hero Oprah Winfrey and Harpo Productions. I figured opening a hair salon/ bookstore was newsworthy, and mine was the first ever in the country. (I actually found that out much later from the people at *Book* magazine.)

Well, it seemed that the idea that someone would put a bookstore in a beauty salon was just so gosh-darned unheard of that *Oxford American* magazine wanted to know more about what we were doing. One day after I had sent the press release, I did follow-up calls, and one of those was to the offices of *Oxford American*. When I explained who I was and why I was calling, the young woman who had answered said, "What? You're doing what, Kathy? Can I call you right back?"

As she hung up I realized that I would probably never hear from her again. To my utter surprise, *Oxford American* did call back, in just a few minutes, and the person on the phone asked if they could send a reporter and a photographer to cover our grand opening.

"Where are you located in Jefferson?" they inquired.

I thought, Now I'm in for it: they think I am in town. Now I had to explain where we were located — how embarrassing!

"Well, you see, my shop is not really in Jefferson. My shop is attached to my house, out in the woods just outside of Jefferson."

I thought that they would most certainly hang up on anyone with the nerve to call them about such a crazy venture, but they exclaimed, "Oh, Kathy, that's even better. Now give us all the particulars and directions." Had I heard them correctly? Were they actually coming to cover my grand opening? First I was stunned, and then I realized I'd better get busy.

> "I got my hair highlighted because I felt some strands were more important than others."
>
> — *Unknown*

Heaven was smiling on me that day, because *Oxford American* sent reporter Carol Dawson, a published author and beauty in her own right, and commissioned Dallas photographer Patricia Richards to take the photos for the feature. All I had wanted was a mention in the magazine and here I was getting a full feature.

To this day, I have a shrine in my Beauty and the Book to *Oxford American* magazine and to John Grisham, who was the publisher at the time. I also give them full credit for putting my shop on the literary map!

Oxford American was clear from the beginning that they understood what Beauty and the Book was about, and I bow to the feet of this magazine, from the publisher down to the copy staff. The magazine told Carol and Patricia to have the whole Beauty and the Book experience — beauty makeover, book makeover (where I match a book to their interests), the works — and, bless their hearts, they jumped right in. The piece, called "Hairdresser to the Authors!" came

199

out in that March/April 2000 issue.

The feature was very complimentary and gave us just the boost we needed. Before long, other newspapers and magazines wrote us up, and after that things just started to snowball. We even got *the* call. The Oxygen Network wanted to do a show featuring us with author Joni Rodgers of *Bald in the Land of Big Hair,* a book about a young woman living in the city of Houston, where your hair is as revered as "Remember the Alamo" is in Texas, who lost all her hair to chemo.

I was thrilled to be on that "Dallas style" show, but I certainly still have to pinch myself to realize we were on it with this little, unknown girl group from Houston called Destiny's Child. Yes, the Kat was on the same show as Beyoncé. I think she has gone on to become a bit more famous than I have!

Visitors to Jefferson started coming to my shop daily to check us out. At the same time, I started getting regular salon customers. I was preaching the gospel according to Kathy to all who would listen: *"Read! Books have the power to change lives!* Can I get a witness? Amen. I said, can you say amen?"

To me, beauty shops happen to be places where you go for a lot more than just get-

> Whatever you do in life, go to the top to get something done. When I opened my shop I e-mailed all the biggies: *Good Morning America,* Harpo Productions, *The Oprah Winfrey Show.* You may not get the attention right away, but if you work hard enough and have a passion for your work, good things happen. Worked for me!

ting your hair done. More than just a service, they are also about community. My shop and others are safe havens where you can spill your guts and not worry that there will be repercussions. It's a place where you go for body, mind, and spirit. The power of touch is sometimes all people need to make them feel better. Show me one woman who doesn't feel better after a good shampooing or a manicure!

Having a combination beauty shop and bookstore enables me to work with my two passions. To some, a job may seem like work. To me, my work is more like play. I have never worked harder in my life. I also have never enjoyed what I have done for a living as much as I do now.

> Having a bad hair day in general? Head to your hearest salon or day spa for a treat. It's like he old L'Oréal ommercial said: "Because I'm worth it!"

The following are reads that show what I mean when I say you can find real beauty in your life, and for me that is the "beauty" in reading:

***My Lost and Found Life* by Melodie Bowsher** Popular high-school senior Ashley has a hot boyfriend and her mother's credit card, and she thinks her life is perfect. Then her mother disappears and by coincidence her mother's company is missing a million dollars. This coming-of-age story, where a young woman's life changes drastically, reminds daughters never to take their mothers for granted.

***Making Waves* by Cassandra King** When I read this book about a young woman who inherits her aunt's house and beauty shop, I thought it was the perfect addition for my shop. What I thought would be a fun little read turned into something more Tennessee Williams–esque, and I adore anything Tennessee Williams.

***Patty Jane's House of Curl* by Lorna Landvik** This is the story of a Minnesota woman who opens a beauty salon that becomes the hangout for the local women's support group. Hmmm, sounds vaguely familiar.

***Hairdo* by Sarah Gilbert** Set in a South Carolina beauty shop, this story about the lives of hairdressers culminates in a road trip to a big-hair competition. It's one of the funniest books I have read. You'll know all these women — and yes, men — who make me love the hair business even more.

***Beauty: The New Basics* by Rona Berg** I highly recommend this book as *the* book for anything to do with beauty. These are the best tips in the world from the former beauty editor for the *New York Times Magazine.* I use something I learned from this book every day.

***God Went to Beauty School* by Cynthia Rylant** This collection of thought-provoking poems will seem sacrilegious to some, but to me Cynthia Rylant is one of the best writers and poets in America today. God doing nails? Well, he always did like hands. You get the picture.

***Facing the Mirror: Older Women and Beauty Shop Culture* by Frida Kerner Furman** This is the true story of Julie's

International Salon, where Jewish women in a large Midwestern city go once a week for salon services. The book addresses what I perceive as groups of women largely neglected in publishing today: elderly, middle-income, and middle-aged. Sounds like the women who frequent my shop.

CHAPTER 7
THE PULPWOOD
QUEENS OF EAST
TEXAS

"Where tiaras are mandatory and reading
good books is the *rule!*"

As I recommended books to my customers
at the shop, they kept coming back to ask
for more. That was all the encouragement I
needed — I decided to start a book club. I
told all my clients, friends, and family that I
was going to start a book club and it was
going to be different. Forget thinking out-
side of the box; there *was* no box. I was
throwing out all the rules except one. It was
going to be *fun!*

Six brave women showed up for that first
meeting. I had put out a fruit-and-cheese
platter and, after I got everyone settled into
comfy chairs with a glass of wine, I began
pitching my idea. I explained, "This is not
going to be your mother's book club."

Everyone perked up a little.

"We will be called the Pulpwood Queens

of East Texas, as pulpwood is the industry in this area. Besides, pulpwood is made into paper and paper into books, but we will not

Kathy,

Although I've always been an avid reader, I'd never joined a book club until I came to the Pulpwood Queens one hot August night two years ago. I never made the time to connect with others who shared the passion to escape between the lines of a well-written novel. Attending my first meeting in Jefferson, I became enthralled at the concept of a very busy woman compiling awesome book titles for her growing club to read. I happily joined this enthusiastic group treated to diverse reading selections *and* live discussions with real authors. It's been an awesome journey with you, Kat! And the rest, as they say, is history. What they might not say, but I will add, is you give a *great* haircut.

With profound thanks for all you share,

Donna, member of the
Pulpwood Queens of East Texas

Pulpwood refers to timber stocks that are cut for paper production. In the logging of mixed forest stands, the better trees are used for saw logs in lumber production, while four types of inferior trees and their components are harvested for pulpwood production. First are trees that are heavily branched lower on the trunk, making poor saw logs. Second are dead and diseased trees. Third are tops and branches cut from trees harvested for saw logs. And fourth are trees too small to be harvested for saw logs. Salvage cuts after forest fires, tornadoes, hurricanes, or other natural disasters often also are used for pulp.

be reading pulp fiction."

Among the women a few titters were heard over the tinkling of their glasses.

"We will crown ourselves the 'beauty within' queens, as we are readers, not fading Southern belles. And who better than a salon professional to show that beauty, real beauty, comes from within? I don't care how dolled up you are. Our motto shall be:

'Where tiaras are mandatory and reading good books is the rule!' "

I could see a deer-in-the-headlights look on their faces. I could sense that this tiara thing might have scared them a bit, but I plunged ahead with my spiel while simultaneously pouring more wine and pushing cheese.

> "A classic is a book that people praise and don't read."
> — *Mark Twain, American humorist, writer, and lecturer*

"We will read one book a month. Our first selection will be *The Divine Secrets of the Ya-Ya Sisterhood,* by Rebecca Wells. I will provide the book-club discussion questions. We are going to read books that don't feel like homework, ones that will entertain, educate, and enlighten us. We will have good food and drink and leave our diets at the door! Did I say there were no rules? Rules are to be broken, for sure, but one must stay in place. You must wear the tiara and you must read the book of the month. Our sole mission is to promote literacy and *get America reading!*"

I jumped up to pour more wine. It was quiet for a minute as I looked around the room to see six nervous smiles. There is a silent issue about women wearing a tiara. No one would dare say this, but most women don't feel they are worthy of wearing one. Only beauty queens should wear tiaras, for goodness' sake, and here I have empowered these women to become queens. I could tell they were hesitant. I had just given these proper Southern ladies permission to cut loose, and it was obvious that they were not at all sure what to make of it.

Since I am partial to jewel tones as well as leopard prints, which I had been collecting for years, I decided it would be fun to make both requirements for our meetings. What little girl hasn't dreamed of wearing a tiara? I began thinking about those tiaras, and about how some of us had not worn one since our high-school proms, if we wore them then. Or how we watched them being placed on the heads of other girls who were being crowned homecoming queen or Miss Texas or Miss America or Miss Universe — or whatever. I thought, Don't all of us gals deserve a crown? Aren't we all queens? You bet we are!

When my speech was over, we all started talking and everyone seemed to relax a bit.

Then before I knew it, it was time for everyone to leave. I announced the time and date of the next meeting and everyone headed out to their cars and homes. I honestly wasn't sure they would show up again.

When the next meeting time rolled around, I cleaned my shop and had it ready to perfection. I prepared enough food for an army of thirsty and starving soldiers. No use expecting the worst, I told myself when seven o'clock was mere minutes away. There was no sign of anyone coming. I kept peeking out the window blinds to watch for book-club attendees and it was a little touch-and-go there for a while, because I didn't see any. Then, suddenly, there came one, then two, then a whole parade of headlights streaming down the black-topped road and pulling up in front of my house. I watched openmouthed as at exactly seven o'clock all these women parked their cars, got out with tiaras in place and casseroles in hand. That night, thirty-five women showed up, way too many to fit downstairs, never mind that I had prepared food for only a fraction of this group. I made a decision.

"Ladies," I announced, "we're going upstairs."

Then, with everyone talking at once, I led them upstairs, out of my perfectly presented shop, to my home that was pure chaos. Dishes in the sink, clothes piled on the dining room table to be folded, books piled on the stairs and on top of the piano. Putting on my housekeeping blinders, I popped the cork on the wine. At that moment I became the fearless leader of the Pulpwood Queens of East Texas.

> "My idea of housework is to sweep the room with a glance."
> — *Unknown*

I had always wanted to be in a book club. My busy sales-rep schedule hadn't allowed it, but I had visited and worked with many reading groups and book clubs. I also had developed some clear ideas about how I wanted my book club to be — fun! — and how I didn't want it to be — stuck-up and exclusive. After literally years and years of talking books with people, I have learned that two things above all others turn people off from reading.

The first turnoff is associating reading with school and homework — with being forced to read books that you may or may not like or to be able to relate to. The second is the aura of elitism that surrounds reading, because some people insist on making a distinction for books that are — choose your adjective — "worthwhile," "literary," "quality," or "important." Well, that is just not me.

I also did not want my book club to be exclusive, so the Pulpwood Queens Book Club is an equal-opportunity book club, all the way. I did not want my book club to be like the one George Moore, the English philosopher, must have had in mind when he said, "A literary movement consists of five or six people who live in the same town and hate each other cordially." I always tell everybody that anyone can join, regardless of race, color, religion, economic background, educational background, age, gender, and whether or not you have had cosmetic enhancements or color-treated hair. Let's face it: a book club is a luxury these days, with time being of the essence. If you are going to take the time to be in one, then why shouldn't you have a good time?

I often think of a book I picked up at the New York airport as I was taking my book club on a literary tour of Europe. That book was *Angry Housewives Eating Bon Bons,* and

My life partner, John, and I first discovered Beauty and the Book on July 4, 2004. We had just finished a busy weekend with guests at our home and business, the Calvert Hotel, Restaurant and Salon in Calvert, Texas. After seeing the shop, all I could think of was Beauty and the Book and the Pulpwood Queens. I e-mailed Kathy to tell her that I would like to start a chapter through my salon, the Calvert Hotel Salon. She suggested starting a Pulpwood Queen and Timber Guy Book Club (for men) chapters. I agreed and started the Pulpwood Queens/Timber Guys of Calvert, Texas. Since starting a Pulpwood Queens chapter, I have made many trips to Jefferson. Last year I attended my first Girlfriend Weekend and was crowned Timber Guy King 2005. Thanks to Kathy and the Pulpwood Queens, I have had the opportunity to meet numerous authors, be a part of various club events, and read works that have enriched my life. I have met many wonderful people that I would not have met had John and I not ventured to Jefferson, Texas, July 4th, 2004. I feel as if my life has

> changed for the best and that I am part of something great and wonderful in promoting literacy — and having a good time doing it.
> — *Nelson Collier, of the Pulpwood Queens/Timber Guys of Calvert, Texas*

the name and cover just jumped out at me from the shelves. I read the synopsis and discovered that the title came from a book-club member's reply when her husband made an unapproving remark about her being in a book club. I could not wait to read about this book club.

In *Angry Housewives Eating Bon Bons* Lorna Landvik introduces the book-club members. They are all neighbors, all very different, and come to know one another through the book club, just as the six women who first came to join our club did not know one another very well. We were brought together because of the books, and are we ever thankful that books introduced us to one another.

We women reign as domestic goddesses, juggling our never-ending car-pool days with the kids, bringing home the bacon and frying it up in the pan. Between motherhood, marriage, kids, career, all our ever time-consuming activities, I sing, "We are woman, w-o-m-a-n — I'll say it again." Women unite! We need to have some girl time, and I can't think of a better way to do that than by having a little wine and discussing a great read.

I have never looked back. These women know all my flaws and lack of organizational and housekeeping skills. Shoot, they've even pitched in, mopping my floors and making up beds for visiting authors and book events. I would never have met these wonderful friends if not for the book club. Most of the women in my chapter don't even live in Jefferson but on one of the surrounding lakes or in nearby towns, and some come from as far away as Shreveport, Louisiana. I cannot imagine my life without them. They mean more to me than anyone can ever know because they are my girlfriends through thick and thin — and, girls, I'm not referring to our waistlines! They have stood by me and I can never thank them enough for helping me on this grassroots effort to get America reading.

At the time of this writing, that official Pulpwood Queens meeting was almost seven years ago. Today, there are Pulpwood

Last December, my friend (and now Co-Queen) Rosanne Hunt and I took a weekend trip to Jefferson with our husbands. Upon our arrival into town, our beaus promptly found the nearest golf course and Rosanne and I were left blissfully alone to shop to our hearts' content for the afternoon. After sampling gallons of free apple cider in little plastic shot glasses in the stores along the main drag, we ventured down a side street and came upon a *very* unique-looking store — Beauty and the Book. Up the steps and through the doors we went.

Every square inch of the place was covered in hot pink and leopard print! Shelves of books lined the far walls, justifying the "Book" portion of the store name. There were also rhinestone-encrusted handbags on hooks, sparkling jewelry, and various other "fun bits" spread throughout the little shop. I noticed that Rosanne had started to salivate. In the adjoining room, an attractive scissors-wielding blonde standing above a chair containing a woman with a wet shag haircut greeted us with a hearty "Come on in, y'all." Aha! There was the "Beauty" part of the title. Never missing a clip, the hairdresser introduced herself as Kathy Patrick, the proprietress. Instantly, we felt as if we had been friends with her forever. As we chatted away, Kathy told us about the Pulpwood Queens Book Club. Since Rosanne and I were both avid readers, we were certainly keen on the notion of being a part of a group of fun women who met to discuss interesting books every month. Face it, women can always find something (or someone) to talk about. It's just our nature. Why not discuss some good books? We figured that if the ones Kathy chose for us to read were anywhere near as interesting as she was, we'd have a ball! The fact that the clubs also promoted literacy was a big plus with us too. So, we agreed to become Co-Queens of the Pulpwood Queens of Grapevine/Trophy Club!

Being Pulpwood Queens has been a hoot and we love it! We couldn't recommend it more!

— *Mary and Rosanne, of the Pulpwood Queens of Grapevine/Trophy Club*

Queens chapters running nationwide, and even in Canada, Scotland, England, Mexico, Italy, and Thailand, with hundreds of members. I don't know how that makes you feel, but it makes me happy — happier than I have ever been in my life.

I started with just six women, and the majority of the membership has spread through word of mouth. A group of girls come into my shop, and the next thing you know they are loaded down with books and cards with my e-mail address to contact when they start their Pulpwood Queens chapter. Women tell their friends and I get a call or an e-mail asking how they, too, can start a Pulpwood Queens chapter. The first thing I do is announce to one woman, "I crown you Head Queen!" She usually laughs in embarrassment as I know she is thinking, How could I have the nerve to be a queen? I will continue to tell her that she is now one of our literacy leaders. I will direct her to my Web site, www.beautyandthebook .com, to read all you would ever need to know to start a chapter.

Of course, being featured on national television doesn't hurt our literacy-crusading book club either, as my phone literally rang off the hook for days after we were featured on *Good Morning America.*

After we were featured in the *Los Angeles Times,* I spent one entire Sunday and Monday in my pajamas taking hundreds of calls asking, "How can I become a Pulpwood Queen?" I am only more than happy to answer this question.

Following are the basic guidelines to starting a Pulpwood Queens chapter.

HOW TO START AND RUN A PULPWOOD QUEENS BOOK CLUB

First of all, crown yourself Queen! You are now Head Queen, which is what I call all the women who run my Pulpwood Queens chapters across the country! Our motto is "Where tiaras are mandatory and reading good books is the *rule!*" Our sole mission is to promote literacy and get America reading! Now that you are the new reigning Head Queen, you must do the following:

1. Decide on a name for your chapter. The Pulpwood Queens of _____. Insert the name of your city or some name of your choice.

2. Find a location to meet each month. That may be members' homes, or your local independent or chain bookstore or library. I have book clubs that meet at bed-and-breakfasts, coffeehouses, a member's

mother's cabin — you name it.

3. Decide on the date, day, and time you would like to meet. This decision might be made after you have your first meeting with a vote among your members.

4. Call all your friends and invite them to join! They don't have to be readers. In fact, I believe a good portion of my membership never really read on a regular basis until after they joined the Pulpwood Queens. We make reading fun and a real pleasure. Isn't that really why most of us read?

5. Don't limit your membership. I have found that even though not all my members come to all the meetings, we have a good-sized hard-core reading group. All I ask is that when they do come, they *do read the book.*

Members will come and go, but you will find that by keeping the membership open you will always have diverse discussion. You will also find that your discussions will always be different and lively.

6. Now, as the leader you must have book-club discussion questions for your group. I provide those for my clubs and also post them on my Web site. Just read the list and everything will fall into place. I would suggest that refreshments and food be served at meetings. While everyone is filling their

plates, make your announcements and then, when everyone settles in, get down to discussing the book. Save other chitchat for later, with dessert or after the meeting.

Then I tell all of those interested in starting a chapter to call me if they have questions after they have read the copy on the Web site.

We normally meet on the second Tuesday of every month and honey, we come decked out. I had a little trouble at first persuading people to dress in a fashion some might find tacky, with hot pink and leopard print — and I learned something important in the process. The women in my book club loved the idea of wearing hot pink tiara-emblazoned T-shirts with leopard-print accessories. I decided we needed official Pulpwood Queens T-shirts, so again, with the help of my girlfriend and artist Carol Hodges we came up with our first official Pulpwood Queens logo.

I had hot pink shirts printed with our book-club name, a tiara logo, and our motto to wear to the meetings. Not long after that, I was in a local clothing store, Amazing Grace, and spied a huge rack of short leopard velvet jackets. I called the club and we nabbed all fifty they had in stock. We

I found out about Kat and Beauty and the Book from a librarian in Shreveport, Louisiana. Kat's reputation precedes her everywhere. The first novel in my Piggly Wiggly series, *Waltzing at the Piggly Wiggly,* seemed like a perfect fit for her wonderful shop, so I e-mailed her about it. Kat and I hit it off and she agreed that *Waltzing,* with its small-town, eccentric Southern characters, was right up her alley. So we arranged for a signing in her shop. Picking up on the theme of my novel — which involves ballroom dancing in a small-town Piggly Wiggly in order to save the store from bankruptcy — Kat asked that I do a bit of ballroom training on her front porch before and after my signing. (It seems her daughter and her boyfriend are interested in mastering the art of ballroom dancing, and I would be showing them the basic steps of the waltz, fox trot, cha-cha, and other dances.) At any rate, a signing/reading/book talk with Kat and Beauty and the Book is indeed like waltzing your way through the book trail.

— *Robert Dalby, author of*
Waltzing at the Piggly Wiggly

took them to another local businessman, Johnny Braley, who embroidered our names on the front in hot pink. Mine says *Kat* on the front, and on the back he embroidered *The Pulpwood Queens of East Texas.*

We now collect rhinestone pins and sequined decals to festoon our jackets that we wear while we are promoting literacy in numerous parades in the area. We ride in the annual Krewe of Hebe Mardi Gras Up River Parade and usually tie their float theme into a book theme. For the past several years we have had a Pulpwood Queens Parade for Literacy, where we literally march with authors around our town's two-block square. I have been grand marshal of parades, tossing combs imprinted with our Web site address or hair-care products and bookmarks. We jokingly tease each other that we'll do just about anything to promote literacy, and parade riding happens to be one of those activities.

Each month we compare rhinestone pins, outfits, and accessories, and that is always a hilarious icebreaker. Several authors have given me pins for my ever-growing pin-encrusted leopard jacket, including Cassandra King's rhinestone pins with SSG on them, standing for *Same Sweet Girls,* the title of her new book.

Since then, as we can't get more of those original leopard-print jackets, we have had no trouble locating other leopard-print or animal-print jackets and coats to buy. As much as we want to have our own unique look, it is fun to have each outfit be a little different to reflect each Pulpwood Queen's personality. Hey, every girl in the group needs an official jacket and some leopard print, but after the tiara and the jacket, they are on their own. Create your own style, is what I say.

You wouldn't *believe* how many different ways there are to wear a leopard print. A lot of women out there secretly love leopard print but never felt comfortable wearing it in public. I have found that women who love leopard print have a certain joie de vivre. Our signature look has given these women license to wear what they have secretly always desired to wear. Now we do it in a big way.

Since we weren't as talented as Carol Burnett at perfecting the Tarzan yell, we settled on "Woo hoo!" as our official cry when author and Pulitzer Prize–winning political cartoonist Doug Marlette did an original design for our Pulpwood Queens T-shirts. He depicted the Pulpwood Queen as a superhero for literacy, and even emblazoned a banner with our yell.

Every month our meeting is a contest to see who can outdo the others with the most outrageous leopard accessory or bling-bling. Once Pam McGregor came dressed entirely in leopard, and then Nancy Williams followed suit. We all shrieked with laughter at their ensembles. I will bet you money that both those girls had on leopard bras and panties that day, too!

Each month we look forward to the latest hot pink or leopard accessory. The best item is announced and we all make a note of where to get it. Once, a local gift shop, Blessings, received a shipment of one-of-a-kind silver rings that you could decorate with beads and charms and, lo and behold, they had leopard beads with tiara charms. We immediately ran to Blessings to get our rings. Later on, they got tiara rings, and we had to have those too!

Nothing we buy is very expensive; in fact, the bigger the bargain, the bigger the thrill. I remember when the local Brookshire's grocery store got a standing display of hot pink Playtex rubber gloves instead of the standard yellow. We had to have them, and what a bargain at something like $4.95.

Discussing *War and Peace* while eating C-rations or *Let Us Eat Cake* while munch-

> When gathering with your girlfriends, leave your diets at the door. Nothing is more important than good conversation and good food.

ing on a celery stick is not my idea of a good time. Instead, I make some mean killer brownies and our Pulpwood Queens Lemon Bang Punch, which has become our signature beverage for Pulpwood Queens events.

Let me just tell anyone who has never been to one of our meetings that we can cook some dishes that are to die for here in the South. In our own club, our Hospitality Hostess Pulpwood Queen Dona Reed, who once operated a bed-and-breakfast with her pastry chef husband, coordinates who brings what and when. Bill delights us constantly with his delectable pastries and desserts, especially his white meringue swans. His Japanese Cloud Cake, served when we read *The Yokota Officers Club* by Sarah Bird, is one of my all-time favorites, but then again his cheesecake, baked with an entire pecan pie on the inside, is sheer ecstasy!

Dona, Hospitality Hostess with the Most-

> ### PULPWOOD QUEENS
> ### LEMON BANG PUNCH
> (For when life hands you a lemon and you want something more satisfying than plain ole lemonade.)
>
> Ingredients:
> 1 bottle lemonade
> 1 bottle ginger ale
> 1 bottle champagne
> Lemon slices and voilà!
>   It serves many thirsty Queens, and during the holidays I add cranberry juice and, of course, more champagne. That recipe has become a part of our growing Pulpwood Queens traditions (more Pulpwood Queens favorite recipes are found in the back of the book).

est, calls all members to notify them of the theme of our next meeting and to note what that member has agreed to bring. Dona was given the prestigious title of Hostess with the Mostest because at one meeting everyone arrived to have our potluck supper and all we had were desserts. Even I cannot make a meal of desserts alone.

Sometimes we make a dish from the book we read that month, like the time we had Sharon Boorstin in for her book *Let Us Eat Cake.* We had to have her Husband Catcher Cake from her book. The funniest dish I ever saw served was when two of the men in our club brought a Chinese noodle salad that was absolutely delicious. We were discussing a book they had recommended: *The Gates of the Alamo,* by Stephen Harrigan. We expected Tex-Mex dishes. I asked them what in the world Chinese food had to do with the Alamo, and Doug replied, "Well, don't you just know that Texas is a great big ole melting pot of people and we have Chinese people in Texas, too?"

I stand informed, laughing all the way.

I once had a male friend who told me I was sexist because I did not let men into my club. Let me assure you that is not the case. I just couldn't get any men to join at first. We could very well have been the Pulpwood Queens and Timber Guys Book Club. As my husband refused to be a Queen, he came up with the term Timber Guy, and the motto the guys have is "We cruise Pulpwood!" One year's special motto was "Got wood?" to go with our band, Woody and the Log O'Rhythms, which plays at our Hair Ball during our annual Girlfriend

Weekend.

Now, these terms may have one connotation to you, but since my husband's whole family has been in the timber business for years, to "cruise pulpwood" means to measure a certain percentage of tree-trunk circumferences in order to estimate the amount of timber you can cut off a parcel of land.

And "Got wood?" comes from the California company Woody's, which makes men's grooming products and has this tagline imprinted on their "boybeater" shirts. This is just another example of how we don't take ourselves too seriously, though I assure you we are damn serious about promoting reading and good books.

The Pulpwood Queens are more than just a book club. We are one another's best friends. We are women who come from every walk of life: We are every age, every size, shape, color, and religion. We have members who wait tables and members who run corporations and banks. We may seem different, but we have one thing in common: We love books!

There is nothing I wouldn't do for one of my Queens, and the feeling flows from one to another, in all directions. We have our own built-in support group. When one

When I first walked through the front door of Beauty and the Book, I wasn't sure what to expect. But in less time than it takes to say, "Hey, girlfriend," Kathy Patrick was giving me a hug. I knew right then that her heart was bigger than her hair — and believe me, her hair is pretty darn big. After asking whether I'd had any trouble finding the place and hearing that I hadn't, she handed me a rhinestone tiara to put on my head and showed me around her shop, a combination beauty salon and bookstore that's almost as zany — and as practical — as Kathy herself. Then we were off to the first of a series of book events she'd arranged for each of the Pulpwood Queens chapters within easy driving distance of Jefferson. I did five talks about my book, which Kathy had chosen as an official Pulpwood Queens Book Club Selection, in three days. Joining us was photographer Rick Vanderpool, who had been commissioned by Kathy to take photographs of her book-club members' tiaras for a Pulpwood

Queens poster print called *The Tiara.*

Every one of these gatherings was a hoot and a half. Even with my tiara on, I was underdressed. (Next time, I'm packing my leopard-print tights!) But the most memorable moment of my trip wasn't on the itinerary. We were heading west on Interstate 20, traveling back to Jefferson after an event in Choudrant, Louisiana, when Kathy said to Rick and me, "Are you guys thirsty?" I nodded yes, thinking she was asking if we wanted some water or soda. But as soon as she pulled off the highway and turned down a dirt road, I realized she had something else in mind. At the end of the path was a shack with a drive-through window and a blackboard menu. No Cokes or Pepsis. Just frozen alcoholic drinks. Rick and I laughed as we read the offerings. It was only two-thirty in the afternoon, but all three of us ordered Mudslides — thick, chocolaty concoctions laced with Kahlúa. A shot of liqueur along with the local color. I'm not a big drinker (nor are Kathy and

member faces a challenge, we face it with
her. Let's face it, between being married or
riding the roads as "soccer" moms, being
involved in our communities, clubs,
churches, and working full-time, we don't
have too much time left for a coffee break
with our next-door neighbor or an afternoon
off at the bridge club. One thing I know for
sure is that women need downtime, girl-
friend time, conversation, and community.

"Good friends are like stars. You don't
always see them but you know they
are always there."
— *Unknown*

We just love to talk, and because our book
club is promoting literacy, it gives us a

legitimate reason to do just that. Does everybody read the book, every time, every month? Doubtful. Are they punished or ridiculed for this infraction? Absolutely not. We punish ourselves enough for not being thin enough, smart enough, rich enough, perfect enough. We receive enough negativity on a daily basis, so why add to the stress? We may be serious about literacy, but we also like to have fun and keep a good sense of humor.

There is nothing in the world I love to do in my shop more than talk about books while styling someone's hair. Here are a few reads that are perfect for the beauty-shop scene and perfect for book clubs, too. And quite a few of these have recipes, so there's all the more reason to choose them for book clubs. And remember, more Pulpwood Queens favorite recipes are found in the back of this book!

***She Flew the Coop: A Novel Concerning Life, Death, Sex, and Recipes in Limoges, Louisiana* by Michael Lee West** No one can write Southern fiction more readable and, yes, funnier than Michael Lee West. This one introduces sixteen-year-old Olive, pregnant by the local preacher. She drinks a poisoned Nehi that sends her into a coma. Meanwhile, the community seems to have only two things on their mind: food and sex.

***Consuming Passions: A Food-Obsessed Life* by Michael Lee West** Here's a book that gives you food, family, Mama, and delicious recipes as well as tips like, "Never serve appetizers, or Better Than Sex Cake or Death by Chocolate at a funeral." Of course those would be perfect for book-club meetings.

***Uncle Bubba's Chicken Wing Fling* by Mitchel Whitington** I give this to men who are reluctant readers. No man can resist the story of Bubba, Aunt Irma, and his bud, Skeeter, and all those chicken-wing recipes for grilling. Women love the book, too. Come visit me and you can meet the author in person — he lives right here in Jefferson, Texas.

***The Persian Pickle Club* by Sandra Dallas** This is a small-town Kansas tale of women who come together over a quilting bee. If you enjoy Fannie Flagg, you'll enjoy a Sandra Dallas book.

***Signs and Wonders* by Philip Gulley** Another book in the Harmony series about the small-town Quaker pastor and his eccentric parishioners. Regardless of your faith or denomination, Philip Gulley's books will make you laugh out loud because you will be sure to know somebody exactly like one of the characters.

Let Us Eat Cake: Adventures in Food and Friendship **by Sharon Boorstin** This is the author's wonderful story of the friends she has had in life and the recipes collected along the way. We served her Husband Catcher Cake when Sharon visited and discussed her book.

Angry Housewives Eating Bon Bons **by Lorna Landvik** Meet the Freesia Court Book Club, five friends who live in the same neighborhood and whose lives unfold through three decades. The big bonus at the beginning of each chapter is a list of each book and author they have read. This book was tailor-made for book clubs.

Eating Heaven **by Jennie Shortridge** A wonderful story about a woman who is not sleekly slim but realistically sized, who also loves to cook. She takes care of her dying uncle, only to really discover herself.

Ruby and the Stargazers: A Fireside, Texas, Novel **by Marci Henna** Fireside is a small Texas town filled with quirky characters, hot gossip, Elvis, and down-home comfort Southern cooking. More satisfying than a big bowl of homemade macaroni and cheese.

***Being Dead Is No Excuse: The Official Southern Ladies Guide to Hosting the Perfect Funeral* by Gayden Metcalfe and Charlotte Hays** Since I live in a town where you are judged by the dish you bring to a funeral, I found this book hilariously funny. We take our funerals and our food dead serious in the South, and we also have a wicked sense of humor.

CHAPTER 8
DIVINE SECRETS OF
THE PULPWOOD
QUEENS SISTERHOOD!

"Friendship is unnecessary, like
philosophy, like art, like the universe itself
(for God did not need to create). It has no
survival value; rather it is one of those
things which give value to survival."
— *C. S. Lewis*

My mother always used to say, "Who needs
friends? Friends cost money."

Call it the stubbornness in me, but I
proceeded to make as many friends as pos-
sible. I have had a lot of friends in my life,
both men and women. But isn't it like they
always say: "Men may come and go in your
life, but your girlfriends will stick by you
through thick and thin"? I am not talking
about your waist here.

Back in my early bookselling days, a galley
of a first-time writer I found pretty amazing
came into my hands. When the book came
out I sold copy after copy, then, all of a sud-

> "It is lack of love for ourselves that inhibits our compassion toward others. If we make friends with ourselves, then there is no obstacle in opening our hearts and minds to others."
>
> — *Unknown*

den, the book went out of print. I called the publisher to inquire when it would be back in print. They informed me that they had decided it had run its course and they wouldn't be printing more copies. I immediately informed them just how many copies I was selling. They thought that was very nice, but they weren't reprinting. They told me it was a regional book. I asked them, "Aren't all books set somewhere?" *To Kill a Mockingbird* could be said to be a regional book. *Midnight in the Garden of Good and Evil* spent about two years on the *New York Times* bestseller list. Tourism in Savannah, Georgia, quadrupled. I believe that was considered a regional book too. The man on the phone said that was nice, but it wasn't their region.

There is something that gets me going more than anything else: the underdog. I was determined to get this book back onto our bookshelves. A sales-rep friend of mine, Katy Stone, stopped by and I practically knocked her down into a seat to tell her about this now out-of-print book.

"That good, Kathy?"

"That good, Katy."

"Can I take this copy with me?"

She was referring to one I had placed in her hands.

"Take it, Katy, and show it to the powers that be."

A few weeks later Katy called me and asked, "How would you like to have the author in your store?"

"Absolutely, Katy. How did that happen?"

"Well, we bought the paperback rights to her first book — the one you recommended — and then she had a second book finished so we bought the rights to that one too."

That first book was *Little Altars Everywhere.* That second book? If you still haven't caught on, that second book was *Divine Secrets of the Ya-Ya Sisterhood,* written by Rebecca Wells.

To say that those books struck a chord with me and now the reading public is to put it mildly. Just about every book-club member everywhere knows about all of Rebecca Wells's books. Her first story is told through a child's eyes, the second book is narrated by a grown daughter, and the third book brings the story full circle to acceptance, love, and understanding. In these books I learned that what happens to us as children gets carried on to the next generation and then the next generation. What we can learn is that the cycles can be broken if we just sit up and pay attention.

> "Friends are God's apology for relations."
> — *Hugh Kingsmill*

I was not surprised when the book was made into a movie. I am a visual reader; as I read, I see the story unfold as a movie in my mind. When the movie was about to release, the director called me from Los Angeles and asked me if I wanted to go to the premiere. Without missing a beat, I inquired, "Can I bring the Pulpwood Queens?"

I'm not joking when I say that this book club is not about me; it's about us. I invite my Queens to go wherever I go, and sometimes they do just that.

"How many Pulpwood Queens would you like to bring?"

It's said that "writers don't *have* to come from a dysfunctional family, but it helps." I think this can also apply to readers. From childhood, books transported me to a kinder, more beautiful world devoid of raging alcoholic fathers, a powerless mother trapped in subservience, and extreme poverty.

For everything lost, something else is given.

Reading gave me the introspective nature to appreciate the mysteries and wonders of man and universe, and taught me that "tragedy given time equals comedy."

So many book clubs are socials serving up lunch, gossip, recipes, and gardening tips. When I inquired about membership in one, I was told, "We're not very literary. Someone did review a book a few months back."

Communicating with other Pulpwood Queens compares to having wandered a barren and alien planet, then suddenly discovering a band of lost sisters of mutual mind and heart who you'd always known were out there.

— *Frances, member of the Pulpwood Queens of Monroe, Louisiana*

"However many can go."

She then told me to check with my members and then to call the contact person for the event, which would be in New Orleans. I am happy to announce that fifteen of my Pulpwood Queens of East Texas, dressed in black sheath dresses, pearls, pumps, and tiaras, attended the screening of *Divine Secrets of the Ya-Ya Sisterhood* in Metairie, Louisiana.

In my previous life, when I was a sales representative, I was on the road a lot. I made friends with bookstore owners and buyers, but my true friends were the books I read at night all alone in motel rooms in Anywhere, USA. Books were my traveling companions, and they were pretty much perfectly suited to my lifestyle. They were quiet, they could be picked up and put down with no complaints, and they were easy to take along on trips. I just loved my books.

> "Good friends, good books, and a sleepy conscience: this is the ideal life."
> — *Mark Twain, American humorist, writer, and lecturer*

When I started Beauty and the Book, it

was the first time I could actually even contemplate taking on something other than kids, husband, home, and work on the road. Suddenly, without the burden of travel and of being away from home for weeks at a time, I had a lot more time on my hands. I had always wanted to be in a book club. But I also always thought if I ever wanted to be in a club that I would like, I would have to start that club myself. You now know the story: six women grew to many, and fast. Somewhere within that first year, I made a startling revelation. I had missed having real friends, friends to share books with, meals with; friends who took the time to listen to me and whom I could listen to in turn. The next thing I knew, we were planning meetings and road trips outside of our bookstore meetings.

After we featured Sweet Potato Queens author Jill Conner Browne, she invited us to attend her Sweet Potato Queens Festival in Jackson, Mississippi. I really love Jill, and she knows how to have fun with her girlfriends better than anybody. We couldn't wait to see what all she had planned at her big weekend festival. They were going to have a contest: Miss Okra, to be exact. Nothing appeals to me more than off-the-wall competitions. Evidently the only criterion for this event was you had to do something really good in a short amount of time. The previous year's winner had won because she could dance really fast. You gotta love a girl who can dance fast!

Now I can do hair, so I came up with this harebrained — sorry for the pun — idea of backcombing and styling a beehive all in the time it took to play the song "Wig," by my favorite party band, the B-52's. The B-52's are a band as American as the Beach Boys, who are known for their dance/party music with such hits as "Love Shack" and "Roam." Kate Pierson, the redheaded lead singer, and Cindy Wilson, the big-haired blonde on vocals, demonstrated that having "big hair" also led to "big-time" media attention. Who was game to go on this girl-friend road-trip adventure?

In the end, two of my book-club members whom I didn't know very well agreed to go. Pam McGregor, who was vice-president of my local bank, agreed to be my assistant, handing me hair spray, combs, pins, and the like. LeTricia Wilbanks, a local beauty who was a stay-at-home mom with two small children, agreed to be my model. We would all dress in leopard print and wear our tiaras. After a practice run in my shop with the girls, we all decided we were more than ready for this girlfriend adventure into the Sweet Potato Queens country of Jackson, Mississippi.

Bright and early, after I set LeTricia's hair on gigantic Velcro rollers, we pulled out of my gravel drive in my damn minivan for a weekend of major girl time. As we talked, the miles and minutes flew by. I had just passed Rustin, Louisiana, on the interstate, when one of the girls said it was time for some bacon, a favorite among the Sweet Potato Queens. I pulled off the interstate after spotting a McDonald's, and we circled around through the drive-through. I was stopped, waiting to place our order, when suddenly I heard a loud bang! I looked back to see who was behind me and a man, hearing the noise, ran out from the McDonald's and yelled at me to open my hood.

Craning to see what was the problem, he hollered, "The piece of plastic that holds your belt on broke."

I always had had a hell of a time with belts.

He walked around to my side of the car, showing me a big black rubber belt that had seen better days. The damn minivan had again lived up to its name. I was fit to be tied to find out that this itsy-bitsy piece of plastic was the only thing that held the belt in place and was now keeping us from our final destination. Where's some good ole American-made steel when you need it?

"What does that mean?" I wailed.

He then proceeded to tell us that we were out of commission until we could get the belt fixed. We would have to have the car towed back to Rustin, since that was where the nearest dealership was located.

> You think you have real true friends. Have something go really, really wrong and you find out who your true friends are right away. In turn, try to be that friend who helps those in need. I try every day to be that kind of friend.

LeTricia and Pam brought my attention to a man who had just parked his tow truck and was walking across the parking lot, getting ready to go into McDonald's. As quick as we could get our leopard accessories out of the way, we jumped out and nabbed him. He looked like we had hit him with a Taser gun, especially as he eyeballed those big red rollers and our leopard-print outfits. The poor man had been on his way to get breakfast for his coworkers when we waylaid his task big time.

We talked him into hooking the van up to his tow truck and giving us a lift to the Chrysler/Plymouth dealership back in Rustin. He informed us we would have to ride in the truck cab with him because it was against the law to ride in a car being towed. As we climbed up in the truck I thought, No way are we going to fit in this cab. I had to straddle the gearshift pole and knob as the other girls were too embarrassed to sit by the driver. Good thing we hadn't eaten yet — we barely fit in the cab. LeTricia had to sit on Pam's lap and she was practically bent over double because her rollers wouldn't let her sit up straight. We almost died laughing as we pulled out. We were all blushing blood red when the man politely changed gears with me in this very unladylike position.

"Excuse me, ma'am. I just . . . er . . . have to . . . pardon me here, move this . . . sorry, sorry . . . gear into place."

This story reminds me of the Ya-Yas riding around in their cars with all the kids in the backseats. Here were some girls who were just having fun and looking for adventure, just like the Pulpwood Queens heading to Jackson, Mississippi. So we were a bit sidetracked. We would get there; we were on a mission.

We drove, chattering all the way, to the nearest Dodge dealership in Rustin. The man who sauntered out told us the service department was closed because it was Saturday. He pointed to the back, telling us we could leave the car in the parking area and drop the keys in the box by the service-department door. They could repair it on Monday. He walked off and just left us standing there, tiaras slightly askew and leopard suitcases at our feet. I declared loud enough for him to hear, "I could have sworn that the South was full of gentlemen, but except for our tow-truck driver, I was wrong."

He never stopped walking or even looked

Take your friends with you. No matter how high you go in life, it's lonely at the top when you arrive all by yourself. I would much rather go arm-in-arm anywhere with my girlfriends. Be a team player. I may have started the Pulpwood Queens, but I hope I have empowered them all to be Queens, and they *rule!*

back. Repairing cars was not his department.

We ran, clicking our high heels, all the way to catch our beloved tow-truck savior to beg him for a lift to a car-rental place. Bless his heart — he did give us that ride.

"Monday?" we wailed at the car-rental place.

Our trip was off. We had called my husband from "Last Resort Rental," but the Jaybird would have none of us renting a car not knowing what it would cost to repair that damn minivan. Jaybird was coming to get us. So we talked this rental-car dude into giving us a ride to the mall while we waited for Jay. Twenty bucks poorer, we were at the mall with our leopard rolling suitcases in tow, big hair in rollers, dressed to the nines in leopard print, and conspicuous as hell. Where to go and what to do for the hours it would take Jaybird to pick us up was the big question. What do most women do when given an afternoon of free time? We shopped, ate, and caught a movie.

We were having a Ya-Ya afternoon. A shop at the mall offered only items in camouflage. There were "camouflage" baby bottles, "camouflage" baby clothes, and real taxidermy animals like bears and deer strategically placed around the store. We stood in disbelief, looking in the window. I thought this place would be perfect for all the hunters in East Texas who festoon their homes with shotgun cases and animal heads.

> "I ask people why they have deer heads up on their walls. They always say it is because it's such a beautiful animal. There you go. I think my mother is attractive, but I have photographs of her."
>
> — *Ellen DeGeneres*

"I would say Louisiana is big into camouflage and hunting," I said as I stared into the store blankly. "Who, may I ask you, would want camouflage baby bedding?"

"Bubba," said Pam, straight-faced, as we burst into hysterics.

I would eat those words a couple years later, when my daughter Madeleine insisted we remove all the Madeline book-character items from her room and change the theme to camo. Madeleine may be named after a little French girl, but she is East Texas through and through. Her biggest thrill was one Christmas when she received a Red Ryder BB gun from Santa and a Crossman

Kathy Patrick is every writer's ideal reader: a down-home, whip-smart, big-hearted (and sometimes big-haired) lover of words and ideas and *fun.* Who else drives her authors 200 miles in a minivan to meet club members, bookstore owners, and disc jockeys, stopping for caffeine infusions and voodoo souvenirs in roadside truck stops along the way? Who else plays the slots with you at the Isle of Capri casino in Bossier City? Who else can leap from weighty social and political matters to comparing Ben Affleck with the underwear model in the Sears catalog in a single conversational bound?

Thanks to Kat, I've gotten to drink champagne for breakfast, ride in my very first parade, dance in crazy costumes with dozens of women I'd never met before who felt like instant friends, sleep in a former railroad car, enjoy the company of writers I've long admired and others I might never have heard of otherwise, wear a tiara on top of my cowboy hat, and pee in the same toilet as Jackson Browne (though not, I might point out, at the same time).She's a force of nature, and to

760 .177 Pellet/BB Repeater Air Rifle from
my daddy.

Only in East Texas will you read in the
letters to Santa printed in the weekly paper,
the *Jefferson Jimplecute:*

Dear Santa,
Please send me for Christmas camou-
flage pants, camouflage shirt and vest,
camo cap, gloves, and a new rifle.

Love,
Beth

We went on to have some orange chicken
at a Chinese take-out place. Then we caught
the movie *Nurse Betty.* Surreal doesn't even
cover that afternoon. I believe that was the
strangest movie I had ever seen. I liked it,
even though I think it could have lost the

graphic violence in the first part.

After we left the dark theater commenting on the bizarre turn of events the day had taken, we rolled our suitcases out of the mall into the bright sunshine. As our eyes adjusted to the glare of the sun and heat, Jaybird pulled up to the curb in front of us right on cue. I was thinking, Man, this day should be a movie. Thelma and Louise had nothing on the Pulpwood Queens. Here was our version of Sam, the father in the Ya-Ya books, ready to save us and take us all home, back to Texas. Divine intervention is all I can say. We were saved.

I really didn't know Pam or LeTricia very well until this trip. Now I can say all we have to do is look at each other and we know that we have bonded as sisters for life. Pam leads a busy life as vice-president of my bank and LeTricia just moved to Tampa, Florida, where she is starting yet another chapter of the Pulpwood Queens.

We may not see each other often, but when we do, look out. We don't have to say anything but "Road trip," and we are on the floor. Have we taken road trips since then? You betcha! None can compare to that one; looking back, it's not about the destination, but the ride — oh, the ride!

> Treasure your friends; they are the diamonds in your tiara of life. Treasure your family; they are the crown jewels.

We never made it to Jackson that day or got to experience Jill and her St. Paddy's Day Sweet Potato Queens Festival. But one day I finally made it to Jackson to visit Jill and the Sweet Potato Queens. Our trip did not go as planned for one minute, but I think we still accomplished our goal. We had fun with our girlfriends and we already had all of Jill's books, which we had read and enjoyed.

The times I share with the Pulpwood Queens are ones I will always treasure. Join a book club, get together with your girlfriends. I think the Ya-Yas had it right, and if you haven't read those books, please do real soon. Be sure, too, to share them with your girlfriends.

Because of the Ya-Yas and how they always took their children wherever they went, I decided to start a "Splinter" chapter of the Pulpwood Queens for my teen daughter and her friends. I encouraged my oldest daughter, Lainie, to run with it, and she did. They choose their own books to read and share.

They also volunteer to do story hours on our front porch for the little kids. Lainie also went with me to the local Head Start pre-school to do a program for Christmas. I was to read *Olive the Other Reindeer* by J. Otto Seibold and Lainie dressed our little Jack Russell dog in reindeer antlers to go as Olive from the book. Since then we have been invited to the church and the elementary school to do the same program. By empowering their Splinter chapter to be literacy leaders, we have been able to double our efforts in promoting literacy.

Since then my youngest daughter, Madeleine, has started a younger-teen chapter called the Pinecones, and they too are now following in our footsteps and the Splinters', in helping to get everyone reading.

After the first year of business, I decided to invite back all the featured authors for the year to a big Pulpwood Queens author reunion party. We called it our Pulpwood Queens Author Extravaganza. But as the book club has grown through the years and it became very clear that mostly women were attending this event, we decided to change the name of our annual Pulpwood Queens convention. We would now call it our Annual Pulpwood Queens "Girlfriend Weekend" Author Extravaganza. And why not expand our annual daylong event into a weekend literary extravaganza where we would call in all the chapters, a Pulpwood Queens reunion party and more? The Pulpwood Queens now host the event and invite every other book club, book lover, and girlfriend. I decided to bring in the big dogs and invite all the worthy authors whose books I could not fit into our twelve-book-a-year selection program.

Each year we have added authors and events to the program. We try to tailor-make our Girlfriend Weekend, featuring the best of authors, books, and literacy, for book

"Remember the time Kaitlyn (my granddaughter) wanted to dress as a Pulpwood Queen for Halloween? She was just five, and when her mother asked her what she wanted to be for Halloween, she said, 'A Pulpwood Queen!' So we dressed her up in my 'Go to Town' hair, a hot pink T-shirt, a pair of black boots, a hot pink feather boa, and (of course) my best and most sparkling tiara. She looked precious."
— *Janie, member of the Pulpwood Queens of East Texas*

I told my publishers that I needed to give the keynote speech at the Girl-friend Weekend, because the Pulp-wood Queens is the largest book club in the world. But the real reason I wanted to go was for a chance to wear a tiara, hang out with other writers, and get down at the Big Hair Ball with a bunch of women who knew how to have a good time.

Kathy claims she encourages lit-eracy, and she does. But I think she does something more, something very important in our electronic-media-obsessed world. She encourages the joy of reading. By her choice of books and her enthusiasm for them, she brings the deep pleasure of literature to people who might have only read for escape and might have missed the delight that reading good writing brings.

— *Loraine Despres, author of*
The Scandalous Summer of
Sissy LeBlanc *and*
The Bad Behavior of Belle Cantrell

clubs and book lovers anywhere. It's our biggest literary party of the year.

That first year of our Girlfriend Weekend was a trip. We had about thirty authors who spoke passionately about their books. We doubled that attendance with women who came from all over Arkansas, Louisiana, and Texas to meet these authors up close and personal. We have had many highlights since then. We've hosted Iris Rainer Dart, author of *Beaches,* who brought the house to their feet with three standing ovations for her presentation on books, literacy, and girlfriends. We've also had local actress Marcia Thomas perform *Texian Woman,* a play she wrote and directed, in her Living Room Theatre, which is actually exactly that: a theater setting in her living room. Her play was based on the actual diaries of Harriet Potter, the first white woman who settled in East Texas on Caddo Lake. We've showcased the book *Love Is a Wild Assault,* by Elithe Kirkland, which has been a local favorite that would make even Scarlett from *Gone With the Wind* say "I swain" about her overcoming adversity in settling in East Texas.

We have also had in attendance author Kathi Kamen Goldmark for her first book, *And My Shoes Keep Walking Back to You.* Kathi helped start Stephen King's Rock Bottom Remainders, which is a band that consists of *New York Times* best-selling authors who have also dreamed of being in a rock band.

> "We play music as well as Metallica writes novels."
> — *Dave Barry, newspaper humor columnist and author*

Kathi and I met on-line when she e-mailed me about her book, and then in person when she dedicated a song to me and the Texas Queens at a Remainders concert held in New York during Book Expo. I brought Kathi back to Jefferson to help me kick off Girlfriend Weekend, and to have her start our own house band with my friends and musicians Dan Garner and Mark Harrell. Richard Bowden, former lead guitarist for Linda Ronstadt, president of the not-for-profit Music City Texas, and Head Timber Guy of my Pulpwood Queens of Linden, Texas, joined the fray. He dubbed our band Woody and the Log O'Rhythms to tie in with the Pulpwood Queens and Timber Guys names. The band would perform on the Saturday evening of my weekend, at an event we call our Hair Ball. The Hair Ball idea came from my designer friend Constance Muller, who had told me they have a Hair Ball down in Houston to raise money for charity. Who better than the Pulpwood Queens to host a Hair Ball to raise money for literacy?

Every year at the Hair Ball we have a theme. This year it is the Pulpwood Queens Go Hollywood; last year it was Out of Africa. We have a costume contest for the best hair and costume and that winner receives the highest award, Pulpwood Queen of the Year. We recently added another category for the best-themed "girl group" because some of the girl groups and chapters just go berserk with their costumes. The Pulpwood Queens of Northwest Houston has wowed the crowd for years with their micromini pink Mylar dresses with faux leopard trim and white platform go-go boots. Beehives are all the rage with that chapter.

At the Hair Ball the hair is piled high and wigs of all shapes and colors abound. Feathers fly from boas, and flashbulbs capture the Vegas-meets-Hollywood-red-carpet Goodwill couture. I encourage the women who attend to get their costumes from the back of their closets, the Salvation Army, or yard sales. Women wear outrageous costumes they would not dare wear anywhere else. I tell the girls every year, anything goes, but be sure you don't wear anything you couldn't tell your friends about in church. We all just die laughing as each woman makes her grand entrance in, say, her strapless ball gown with leopard bra straps showing. Once a girl came in a red (or "read"), white, and black NASCAR gown that had actual race cars running around the bottom striped ruffle with NASCAR emblems and checkered flags. One Hair Ball Pulpwood Queen of the Year had actually had her husband, a wire artist, wire her waist-length hair into a tree with lights. You actually have to experience the Hair Ball to believe it!

Woody and the Log O'Rhythms play dance music, and some of the authors in attendance join the band for cameos during the evening. Author Kathi Kamen Goldmark always brings the house down with her rendition of a song she wrote involving a woman who has become invisible to the opposite sex in middle age. Everybody comes with all their girlfriends and we just all dance like crazy as a big group. No partner dancing here. We are celebrating our joy of girlfriends and literacy with *big hair, big time!*

Every year as I collapse into a chair to catch my breath, I look out at the throng of

There's something magical about Girl-friend Weekend. It's hard to explain. You have to surrender. Take the plunge. It's one of those "check your ego at the door" things. Whoever you think you are when you arrive won't be who you are when you leave. Friday night at the local honky-tonk, I found myself singing karaoke to Shania Twain's "Man! I Feel Like a Woman!" with a bunch of women I'd never seen before in my life. Sunday morning, I even took the purple rhinestone tiara out of the gift packet I'd received upon arrival. Now I have never been a tiara-wearing type, so this was a real stretch. After much maneuvering in front of the mirror, I finally placed it upside down on top of my blue knit skullcap. The fake-pearl and rhinestone top hanging down between my eyebrows gave off a certain Hindu vibe that felt right. So I wore it that way to the Pulpwood Queens brunch. And continued wear-ing it that way for the three-hour ride back to the airport with my friend Kathi. Our spirits were high as we laughed and sang and talked about the week-end. We never even stopped to eat.

> But we did write a song.
> — *Marshall Chapman, songwriter of*
> *"I Want My Hair Considered"*
> *and author of*
> Goodbye, Little Rock and Roller

women, some who probably hadn't danced this merrily and uninhibitedly for over twenty years, and my eyes fill with tears of joy. The Hair Ball is our biggest party of the year and everybody, I mean everybody, has a good time. Authors are dancing, book-club members are dancing, and girlfriends are dancing on the floor to songs like "Wooly Bully" or "Boogie Woogie Bugle Boy of Company B." There is something for everybody at the Hair Ball.

Each year we join in a parade or two, including Jefferson's Krewe of Hebe Mardi Gras Up River Parade. Our float presence began with the Spirit of America theme, and we did a float titled the Pulpwood Queens DO Miss America. Last year the theme was Gypsies, Tramps, and Thieves, but whatever the theme, we parade for literacy.

We have had authors such as Jill Conner Browne ride on our float, and some authors

When you show up for a book event and the first thing your hostess says to you is "Do you like pah?" (meaning "There's a little place in Jefferson where they make the best coconut pie in the world and we're going to get some right now"), you know it's not going to be a typical book signing. From the moment I hit Jefferson, Texas, for the first annual Pulpwood Queens Girlfriend Weekend, I was anointed as one of the girlfriends. So much about that weekend was, um, unusual: the dress code, tiaras and leopard print (I'm the kind of person who will pack rhinestone earrings on a camping trip in case I'm invited to a surprise prom); the fabulous women like Margie Dilday, who immediately clucked and cooed over my impending divorce, and the improbably named Dona Reed, who confided (at age sixty-something) that joining the Pulpwood Queens helped give her the courage to be herself and stop worrying about what others thought of her. The best of all was my Pulpwood Queen makeover, accomplished in Kathy's beauty-parlor chair and videotaped while she interviewed me. What woman can assume the position in that chair and not tell all? I left Jefferson, Texas, with new friends, all of whom had read my book, and a hairdo that allowed me to fit right in at the Dallas/Fort Worth airport. Whoever said literary people are stuffy has never visited the Pulpwood Queens.

— *Kathi Kamen Goldmark, author of* And My Shoes Keep Walking Back to You

ride in convertibles with us, including Marlyn Schwartz, former columnist for the *Dallas Morning News* and author of *The Southern Belle Primer: Or Why Princess Margaret Will Never Be a Kappa Kappa Gamma,* and Jefferson's own Fred McKenzie, my fellow eighty-eight-years-young bookseller.

We have a blastie blast putting together these floats and costumes, all in the name of literacy. Pulpwood Queen Janie Ready wrote this poem about the float our Harleton, Texas, teachers' Pulpwood Queens put together for us one year.

THE HARLETON 5

By Janie Ready

Was a few weeks before Mardi Gras
And up in Harleton Land
Five little Pulpwood Queens were busy
    hatch'n out a plan.
There was Kathy Bartuska as the leader,
Wendy and Janie, too.
And don't forget Judy and her sister, Cathy
    Lou.

They would build a float
As only they could do.
Yes! Yes!
With colors of hot pink and purple and
    leopard print too!
Their creation would be outstanding,
    fabulous — simply divine.
On this float they would stun everyone at
    Mardi Gras Parade Time.

And stun they did by winning top prize!
This team of Queenly Girlfriends
Know as "The Harleton 5."

Every year the Girlfriend Weekend is such
a success that everyone who comes is back
the next year with even more girlfriends.
Authors return, bringing even more authors,

and the word continues to get out. Now, authors and attendees travel from all over the United States to Girlfriend Weekend. We look forward to this event every year, as it is always held the third weekend in January here in East Texas.

I have been on a roller-coaster ride ever since we started taking road trips together and meeting up for Girlfriend Weekend. I have found that we can get through a lot of things with our girlfriends. I have taken a lot of trips through the years, but the ones with my girlfriends are definitely ones that I will never forget.

> "Friendship is born at that one moment when one person says to another, 'What! You too? I thought I was the only one.' "
>
> — C. S. Lewis

Great books will inspire you to get together with your girlfriends and maybe take a road trip, whether armchair or otherwise. Here are a few that are perfect to share.

The Last Girls by Lee Smith Who could not love a bunch of college girls who built a raft to go down the Mississippi River? In this book it's thirty-five years fast-forward, and they reunite for a cruise. This book made me and all my book-club members want to do the same.

The Saving Graces: A Novel by Patricia Gaffney Four women in Washington, D.C., have been meeting once a week for ten years. This is the perfect read for women who want to meet and discuss themes like friendship, men, careers, babies, and — yes — even death.

The Florabama Ladies' Auxiliary & Sewing Circle by Lois Battle Fifty-year-old, privileged Bonnie has to start over, as her husband has left her not only for another woman but also in the throes of bankruptcy. Never having worked in her life, she is hired by the local college to run a program for displaced homemakers. My club really related to this woman's story, and we found it made for great discussion.

Cover the Butter by Carrie Kabak When Kate, a woman in her forties, arrives home in Wales, she discovers that her son has trashed her house while she was away. Kate sets forth to put her desperate-housewife life back in order.

The Future Homemakers of America by Laurie Graham This story follows the lives of six women thrown together as military wives from World War II to the 1990s. We see them go from England to America, and on the final page you will feel you know these women as well as your own friends.

***The Same Sweet Girls* by Cassandra King** Cassandra has been getting together every year with the same sweet girls since college, and, drawing on this, she wrote a book about female friendship. Cassandra is also a three-time Pulpwood Queens Book Club selection author, so to say we love her books is a complete understatement.

***Gladys on the Go* by Kelly Povo and Phyllis Root** This short, inspirational picture book for women is the one I pull from my shelves to read aloud to girl groups when they visit my shop! It has funny photos and is hysterically inspiring to girlfriends!

***Wild Women and Books: Bibliophiles, Bluestockings, and Prolific Pens* by Brenda Knight** This is *the book* for book-club members who are addicted to authors and books as much as I am.

CHAPTER 9
BOWING TO THE FEET
OF OUR QUEEN!

"I crown you all queens, for you *rule* by
being readers!"
— *Kathy L. Patrick, founder of the
Pulpwood Queens Book Club*

I first met Joyce when she came to Beauty
and the Book for a routine wash and blow-
dry. The last time I saw her was at the Home
Sweet Home Funeral Home. I was a ner-
vous wreck that day. But I had promised to
make Joyce look like a Queen, and, where I
come from, a promise is a promise.

Joyce had had cancer. It was pancreatic
cancer, to be exact, but Joyce never came
right out and said so herself; I only knew
because her family and friends told me. It
was in Joyce's nature never to talk about
unhappy things when we were together. I
knew she'd been through chemotherapy,
because the first time I met her she ex-
plained that her hair had just grown back.

It was a faded strawberry blond, all the same tone, definitely from a bottle, without one iota of shine. She wore it just below her chin in a layered, feathered-back style. It looked like a big puff of butterscotch cotton candy.

When I think of Joyce, I think of Scarlett in *Gone With the Wind.* That undeniable spirit of Scarlett's — my way or no way — fit Joyce to a T. Like Scarlett, Joyce had a manner that made you believe that hers was the best way.

Joyce's hair wasn't the first thing you noticed about her. Instead you would see her striking features, like that first image of Scarlett at the garden party in the opening of *Gone With the Wind.* Joyce wore very dramatic makeup, and you just sat up and paid attention when she entered a room. Though she looked nothing like Scarlett, I thought she was the spitting image of one of my mother's favorites, actress Agnes Moorehead. Joyce's eyebrows extended straight up toward her temples, just like Endora's, the Agnes Moorehead character in the television show *Bewitched.* Others thought she looked just like actress Shirley MacLaine.

Although Joyce was a small woman, she was a large presence in my life and in the lives of the Pulpwood Queens. By the time we met, the chemo had already taken its toll. She seemed frail — until you saw her with Skeet.

> "I think of life itself as a wonderful play that I've written for myself, and so my purpose is to have the utmost fun playing my part."
> — *Shirley MacLaine*

Joyce was a ventriloquist, and a damn good one in a field that has fallen by the wayside. Skeet was her vent or, as those of us not in the ventriloquist business would say, her "dummy." Skeet was a collector's item, a throwback to the time when vents were actually carved from a single piece of wood and were quite valuable.

Toward the end of her illness, Joyce had her good days and her bad ones, but when she performed with Skeet, she was transformed. Sitting on a dark stage, illuminated by a single spotlight, Joyce would flip the lock of his battered black leather suitcase and gently lift Skeet up and onto her lap, burying her hands in the folds of his little black suit. "Say hello to everybody, Skeet," she would whisper in a shaky voice. Suddenly, this little man, with an attitude of pure mischief, sprang to life. "Hello, dummies!" he would shout in an amazingly chipper falsetto that seemed to blast through the room at a hundred decibels.

In beauty school they teach you that there are three things you never, never, *never* discuss with clients. They are religion, politics, and sex. Honey, Joyce and I talked about all three from day one. Joyce was a natural entertainer and, boy, could she tell a story. I would be laughing my head off at something she said one minute, and tears would be rolling down my face the next.

Joyce was deeply spiritual, and one of the most positive people I have ever met. She was a Southern Baptist and I am a Methodist, but she had a way of speaking the truth that was so profound it emphasized our similarities, not our differences. Listening to her, I felt uplifted, inspired, and at peace — all at the same time. I think Joyce's faith helped get her through her illness, and I know my conversations with her helped me affirm my own faith.

After that first day, Joyce had a standing appointment for a shampoo, blow-dry, and manicure. I scheduled her as my last appointment of the day so we'd have plenty of time to talk. If she came in earlier, I would end up playing catch-up with my other clients for the rest of the day, since we tended to talk too much. That's how we spent most of our time together: after hours in the shop, just the two of us talking, me drinking my super-strong coffee and fussing with her hair or painting her nails while she sipped her organic green tea.

Like me, Joyce loved clothes. Her signature style was a mix of vintage, Victorian, Southern belle, and gypsy. She had even bought a pattern of Scarlett O'Hara's garden-party dress that we were going to make into a costume for her for a Pulp Fashion show. I think Joyce would have loved to have worn every costume of Scarlett's.

We shared a passion for big jewelry and animal prints. She was forever presenting me with little finds — leopard-print socks or an antique African bracelet — and I did the same for her. It tickled me to no end if I could find something original and unique to give her. If an event called for a costume, we were the first to dive right into finding the perfect outfit. But what really cemented our friendship was that Joyce was a reader. If someone wants to talk books, my goodness, they are a friend of mine for life.

Like Scarlett, Joyce never did anything

halfway. When she joined the Pulpwood Queens reading group, she became the model Pulpwood Queen. She rolled up her sleeves and went to work. No time for self-pity; there was work to be done.

> "There is no higher religion than human service. To work for the common good is the greatest creed."
> — *Woodrow Wilson, twenty-eighth president of the United States*

She came to every meeting dressed in full Pulpwood Queens regalia, including the official Pulpwood Queens rhinestone leopard jacket emblazoned with her name in cursive on the front left-hand side. If Scarlett O'Hara had been a Pulpwood Queen, she would have worn her jacket to every meeting, too. Joyce never shied away from giving her opinion about the book we were reading, and this always guaranteed lively discussions.

Joyce's all-time favorite book was *Gone With the Wind.* Joyce saw herself as a modern-day Scarlett, misunderstood and forever searching for her Rhett Butler. She, too, loved fine things and adored being the center of attention. But even more, Joyce had Scarlett's stubborn determination. It's a good thing, too. She needed it in her battle with cancer.

We all knew Joyce was going to die, but we didn't dwell on it. Instead, the Pulpwood Queens banded together to provide support wherever it was needed. Joyce would not tolerate pity, but if you wanted to help, Lord, yes, you were welcomed with open arms — and plenty of instruction on just what to do.

When she was too sick to cook, we'd volunteer to make supper for her and her husband, Doyle. Her favorite was roast beef with fresh green beans made by Pulpwood Queen Elizabeth Stokes, with Pulpwood Queen Mary Kay Rex's homemade banana pudding for dessert. (From me, she was more likely to get one of the pineapple shakes she adored from the Sonic drive-in.)

If one of us stopped by her house in the morning to see what she needed that day, she might hand over a shopping list.

"And — oh, by the way, darlin', my cousins from New Orleans are in town. Do you think you could prepare dinner for all of them, too?" She would practically purr as she said it.

"Sugar," she explained if someone teased her about it, "you can catch more flies with honey than with vinegar."

Yes, she could be, as my daughter Lainie said, commanding. I don't know if it was because we knew her time with us was limited or because she had an inner strength and wisdom that made us want to be around her — she claimed to have a special relationship with Jesus Christ, and you won't hear me denying it — but we waited on her hand and foot. I did not mind one bit.

I think it worked the other way, too. Joyce took strength from the Pulpwood Queens. No matter how bad she felt, she never missed a meeting. I know it sounds crazy, but she could walk in looking like she was at death's front door and, by the end of the meeting, the color had returned to her cheeks. She'd be beaming.

Through Joyce, I saw firsthand how cancer ravages a human being. Her hair and nails grew more brittle and her step less sure every week. Right up to the end she refused to give in to fear or self-pity. She believed that as long as she kept telling herself "I can beat this thing," she, not her illness, would be in control. She never talked about being sick; the C-word never crossed her coral-painted lips. If cancer was going to kill her, she wasn't interested.

"You never know when it's your time to

go," she would say. "Land sakes, I might get run over by a truck tomorrow!"

All the same, Joyce was *not* one to let herself be caught unprepared. That's why, toward the very end, she started dropping hints about the kind of funeral she wanted. She tried to play this down, but what she was doing was obvious.

"We're all going to die someday." She'd laugh. "Best make those plans early."

Truth is, she knew exactly what she wanted, and she was adamant. Joyce Jackson Futch would go out in style. She planned her own funeral, which turned out to be one that would have made Scarlett O'Hara swoon with envy.

Joyce grew up in New Orleans and had always loved the way funerals in that town weren't somber affairs, but a joyful celebration of the deceased's life. She wanted a New Orleans jazz band to accompany her coffin as it made its way from the First Baptist Church on Polk Street to the Oak Forest Cemetery. She had a floral designer extraordinaire, our friend Dale Vaughn, all lined up to procure the band. After the burial, she wanted everyone to march back to the church, singing, dancing, and rejoicing in the streets of Jefferson.

> "I would rather be ashes than dust! I would rather that my spark should burn out in a brilliant blaze than it should be stifled by dry-rot. I would rather be a superb meteor, every atom of me in magnificent glow, than a sleepy and permanent plant. The function of man is to live, not to exist. I shall not waste my days trying to prolong them. I shall use my time."
> — *Jack London, American short-story writer and novelist*

She wanted to be buried in one of her stage outfits, a full-length, crimson and gold sequined number dripping with beads and crystal. I had seen her perform in this gown many times. It screamed, "I am the star of this show — look at me!" In death as in life, Joyce *was* a star. Naturally, she would meet her maker wearing her Pulpwood Queens tiara.

One Saturday afternoon in August, I was painting Joyce's nails with her favorite color, a copper rose called Whatever! We were chattering away about something — I can't remember what — when suddenly she got quiet. I kept on working, but I could feel her looking at me. Then she said, her voice soft and low, "Kathy, will you do my hair and makeup for me when I die?"

Oh, I had just known this question was coming. I'd already thought of every excuse in the book. Suddenly, I felt like Scarlett when Miss Melanie asked her to take care of her dear, dear Ashley, torn to pieces. I can't stand funerals. . . . I don't think I could handle it. . . . Joyce, you know how I always faint at the sight of blood; imagine what'd happen around a — Oh no, darlin' — they use a special kind of makeup, I'm not qualified.

Now, to my own amazement, I didn't say any of those things. Instead, I looked her in the eye and said, "Yes. I'll do it."

Joyce was my friend; I couldn't turn her down. "But," I quickly added, "you have to tell me exactly what you want me to do. I want you to look like a Queen."

So after her manicure, she showed me the colors she wanted, told me her favorite brand of foundation, eyebrow pencil, and the rest.

"Just go by my house and the makeup kit will be ready," she said. We practiced a couple of times and that was that. We never mentioned the makeup — or dying — again.

Looking back, I realize that Joyce's illness

escalated quickly after that, but at the time everything seemed to happen in slow motion.

In October, I had to go to the funeral of a friend's husband across town at the Christ Episcopal Church. On the way, I made a stop in Joyce's neighborhood. Her friend Leska had picked up a Charlie McCarthy doll at a garage sale. Thinking I might want it for Lainie, who had been taking ventriloquist lessons from Joyce for the past year, she asked me to come have a look-see. Leska's find turned out to be a pull-string doll, not a real vent, so I couldn't use it. I had a little time to kill before the funeral started, so Leska and I chatted for a while in her driveway. Turning to leave, I noticed several cars over at Joyce's house.

"Looks like Joyce has company today," I said to Leska.

Isn't it just always the way? Something like that should have been such an obvious clue. It didn't even register. I left Leska's and didn't give it a second thought.

I pulled into my driveway right about suppertime, parked, and was pulling down the garage door when the phone in the house started ringing. I ran up the steps in the garage to enter the kitchen and answer the phone. The voice on the other end was

Chere, Joyce's daughter, calling to give me the news.

As much as we prepare ourselves for death, we're never really ready for it when it happens. The loss just feels too great, the knowing that "I'll never talk to her again" so very final. I loved Joyce; I tried to let her know often, and I think she did know. Still, you can't help but think about all the times you felt it but didn't say it, and wonder, Did I tell her often enough how much she meant to me? It's just about a guarantee that you'll feel you came up short. If there is one thing I learned from Joyce's death, it's that we should tell our loved ones that we love them *every day.* We should cherish each earthly moment we have together.

With a shiver I remembered my promise to Joyce. I also remembered how Scarlett mourned her own mother. All the dread returned, and I wasn't sure I could really go through with it. Then I got the call from the funeral home: Could I come Sunday evening around seven? Yes, of course, I'd be there.

There was no way I wouldn't go through with it, but I knew I couldn't do it alone. So the minute I hung up the phone, I was on the line again, calling Pulpwood Queen Kay Brookshire. Kay is a rock-solid friend who had also been close to Joyce, and I knew if anyone would understand my dilemma, it would be Kay. In a crisis, Kay is the person who always knows exactly what to do and somehow manages to stay calm. She's nothing at all like me, since I was a nervous wreck just *thinking* about walking into that funeral home. I knew Kay would be there for me. That's how it is with the Pulpwood Queens: we support each other big time.

To say that I felt sick on the drive to Home Sweet Home Funeral Home would be putting it mildly.

"I don't think I can do it, Kay," I kept saying. "I just know I'm going to lose my breakfast or faint, or both. I'm serious — this could get ugly."

Kay just kept reassuring me, "Joyce is in heaven, Kathy. The body is only a shell. You can do this. You'll be fine."

I was grateful for her kind words, but my stomach was still churning — not just with butterflies, but doing somersaults, cartwheels, and back flips — as we turned into the parking lot. I felt a little faint as we got out of the car. Kay gave me a big hug and we walked arm in arm to the side entrance, where the funeral-home director, Mr. Marion Clark, was waiting.

"Hey, Marion. I'm here to do Joyce's hair and makeup," I whispered.

"Yes, it's so good of you to do this, Kathy. Joyce is all ready to go," said Marion, leading us inside. He was all business.

"This way," he said, turning to the left as Kay and I followed a few steps behind. He reminded me of a waiter in a fancy restaurant. He spoke in a normal tone of voice, but in the eerie quiet he might as well have been yelling at the top of his lungs.

"Kay, why are we whispering?" I whispered.

"Because we think they're sleeping?" she answered.

"Right. I guess we don't have to whisper," I said in a slightly louder voice. "Who's listening anyway?"

We giggled self-consciously; our levity felt very out of place.

At the end of a short hall, Marion stopped at a door on the left. "Here we are," he said breezily.

Already? I don't know what I expected — maybe that we'd walk down long hallways with endless twists and turns before descending to a cold, dark basement full of dead bodies laid out on operating tables and mad scientists wielding strange instruments over them, like something out of Mary Shelley's *Frankenstein.* Actually, we were just a few feet from the home's main entrance. I held my breath as Marion swung the door open. There, in a room about the size of a large walk-in closet, with walls covered in flowered wallpaper and nary a cobweb or scalpel in sight, was my friend, lying on a stainless-steel table with a white sheet draped over her body. For all the terror I'd felt on the drive over, now that I was standing not three feet away from Joyce's body, suddenly I was not afraid anymore. I felt calm and certain that Joyce was with me, taking care of business in spirit. I was filled with a sense of the importance of my mission. I *can* do this, I thought. I *will.*

Next to the table were cabinets and a long countertop that ran the length of the room, just like in a kitchen. I got out my hairdressing tools and Joyce's makeup kit, which we'd picked up from Chere on the way over, and set them down on the counter. Then I stepped closer to the table and pulled back the sheet to take a good look at Joyce.

Her hair had been washed and dried and her neck was supported by a Styrofoam block. Her stiff body was light as a feather, but when I gently touched her arm with my finger, it felt greasy. I asked Marion why, and he explained that they put oil on the body to prevent dehydration from the embalming fluid. I'll have to wipe it off when I put on her makeup, I thought. I had intended to do a manicure, but when I picked up Joyce's hand and saw that there wasn't much play in her fingers, I decided just to add a coat of polish, the two-toned gold and coppery red she had asked for, and leave it at that. Next, I took my curling iron and started curling her hair.

Now this was interesting. Have you ever tried to comb someone's hair when she's lying down? It's awkward at best, and this was worse. Joyce's hair was so limp that even after I curled it, it kept falling away from her face. I had to spray the living daylights out of it with hair spray to get it to stay put.

As I worked, I hummed a little and talked to Joyce as if she were in the shop, and, for some reason, I had an insane desire to describe everything I was doing to Kay and Marion, like a TV chef. Marion piped in with questions every now and again, and he was actually taking notes! Even Kay asked a question or two and said to be sure to tell her if there was anything she could do to help. Anyone walking by would have thought there was a party going on, such a fine old time we were having at this makeover. I know it sounds perverse that we were having such a good time and all, but I am convinced Joyce was in that room with me. And that's just how she wanted it.

Next came her makeup. Joyce's skin looked slightly jaundiced, and I was going to have to do a little correcting. That shiny oil definitely had to go. I wiped it off gently with cotton balls, applied a rose base to offset the yellow, and then covered it with her regular foundation, a peaches-and-cream color that matched her complexion.

"She looks so natural and lifelike!" Marion exclaimed.

Surprised that he was surprised, I asked, "Well, who usually does the hair and makeup for the deceased?"

"I do," he replied.

"What do you use?" I asked, genuinely curious.

Marion bought his makeup from a funeral supply company, a heavy paste foundation — the kind that should have gone out with the dark ages. The main virtue was that it covered everything, like a mask. I added a little blush, some on the apples of Joyce's cheeks, and then brushed on up into her hairline to add a bit of color.

"What about eyeliner, mascara, eye shadow, and eyebrow pencil?" I cried, incredulous.

"Never use it," said Marion.

Now I understood why I've always hated viewing dead bodies at funerals! Bad makeup job. Hey, I could make a *fortune* teaching these professionals a few tricks of the trade, I thought, which made me laugh to myself. I had the feeling Joyce was laughing, too.

Then I noticed that one of Joyce's eyes looked like it might be coming open.

"Nothing a little glue won't fix," said Marion, taking a tube of glue and tweezers out of the cupboard. Then, holding the eyelid down with the tweezers, he proceeded to glue it shut, just as nonchalant as you please. My face probably showed my shock.

"Happens all the time," said Marion, returning his tools to the cupboard.

Next I drew on Joyce's "Endora" eyebrows. Now, for a little mascara, eye shadow, and a tad more blush.

Lipstick came last. Was that a piece of cotton coming out of her mouth? I brought Marion's attention to the wisp of cotton that was just barely showing in one corner.

"No problem," said Marion. "Let me have the lipstick you are going to use."

He proceeded to take a little piece off the extended lipstick and mixed it with what looked like clear caulking. He pulled out the piece of cotton with the tweezers and ran a bead of the colored gel across Joyce's lip line, then, using a brush, he blended it in the line to seal her lips.

> "Whenever you go to town, Katsoup, don't forget that you're not dressed unless you have on your lipstick."
> — *Helen Marie Kelsey Maloney, Kathy L. Patrick's grandmother*

"There, now her mouth won't open. The cheeks always sink in after death so we stuff the mouth with cotton. A little bit was just coming out."

I was amazed by this procedure. Then I noticed that Joyce's mouth sagged a little on one side, so I dabbed a little concealer over the lip line, covered it with foundation, and, using a lip pencil, redrew her lip line, raising it to make it look more natural. I filled in the lip pencil with her lipstick. There, all done. I stood back a little to assess the results.

I smiled. Instead of wearing a grimace, Joyce looked serene, like a Queen. Letting out a long sigh, I began packing up my belongings as Kay said her good-byes to

Joyce and Marion. Then I took one last look at my friend.

"Good-bye, Joyce. I'll see you later, girl-friend," I whispered, yet again for no known reason. "You look beautiful."

> "Only happy people can learn. Only happy people can teach. Our religion should put a sparkle in our eyes and a tone in our voice, and a spring in our step that bears witness to our faith and confidence in the goodness of God."
> — *Unknown*

Joyce looked to Scarlett O'Hara as her inspiration. I looked to her for mine. I reread *Gone With the Wind* not too long ago, and all I could think of was how Joyce's spirit could live on like Scarlett's. Though the film was wonderful, please read the book and please remember my friend Joyce. These women are our heroes who show the great determination and inspiration of the American woman's spirit. The following books include characters who might inspire you.

***Love Is a Wild Assault* by Elithe Hamilton Kirkland** This book is based upon the true diaries of Harriet Potter, the first white woman to settle on Caddo Lake near Jefferson, Texas. I tell everyone Scarlett had nothing on Harriet for the adversities she had to overcome. A must-read for anyone who loved *Gone With the Wind.*

Storyville **by Lois Battle** The story is about two women, one who as a young girl is seduced and thrown into a life of prostitution in the New Orleans red light district in the late 1800s. You won't be able to put this book down.

True Women **by Janice Woods Windle** An unusual and intriguing mix of family memoir and historical novel, this is a *Gone With the Wind*–style story involving brave Texas women, set in the mid 1800s to 1940s.

Private Altars **by Katherine Mosby** Poet turned writer Katherine Mosby crafts the story of a young New Yorker who marries a West Virginian during the Depression and becomes an outcast in their society. This is another story that confirms the belief that women are much stronger than they look.

Having Our Say: The Delany Sisters' First 100 Years **by Sarah L. and A. Elizabeth Delany with Amy Hill Hearth** Two sisters born to a former slave whose owners broke the law by teaching him to read continue that tradition of teaching and learning by telling their story. Since both sisters are over 100 years old, this story is one of the best oral histories about what it was like for African Americans who pioneered as professional career women.

Long Live the Queen! Posthumously, we crowned Joyce Jackson Futch Pulpwood Queen of the Year, 2003. Her throne sat empty on our floats in parades throughout the year. We will never forget Joyce, our Scarlett O'Hara, and we bow as always to the feet of our Queen.

For those who wish to visit Joyce's gravesite, she is buried in the Oak Forest Cemetery here in Jefferson. We are also taking donations to buy her a tombstone, as all she has now to remind us of where she is buried is a little metal marker, and for a scholarship fund that we hope to establish in her memory. Those checks may be made payable to:

Joyce Jackson Futch Memorial Fund
c/o First National Bank of Hughes Springs
Jefferson, Texas 75657

CHAPTER 10
HERE SHE COMES,
MISS AMERICA!

"It's very surreal. When we were little we would watch the pageant as a family, and my sister and I would go into the utility room and play Miss America in our heels."
— *Jennifer Berry, Miss America 2006*

I am seriously thinking of investing in a leopard carpet, one that I can roll out for all royalty who walk up into my shop. Customers, clients, visitors, authors, Pulpwood Queens, Timber Guys, dignitaries, anyone walking or riding, I want all to be included so they can feel just like Miss America!

A basket of loaner crowns and tiaras would be placed at the door, and a scepter to carry would complete my picture. I have even made ready a Queenly or Kingly cape to wear as the King or Queen graces me with a visit. The cape is black and trimmed in hot pink fur. (Wal-Mart has amazing hot pink bath mats that have certainly come in handy for me.) A dozen long-stemmed roses would add the final touch, and I know just the ones to get from the Dollar Store. They'll last forever and look so lifelike.

Not a day goes by when I don't experience a Queenly moment. Someone who makes me feel more royal makes me thankful I'm alive. I recently read *Miss America Pie* by Margaret Sartor, and that book brought back to me all my teenage angst. The book consists of Margaret Sartor's diaries from seventh grade to high-school graduation and is a slice of my time: the 1970s. The book reminded me again that I was not the only one trying to find my place in the world.

When I finished the book, I immediately e-mailed the author this letter:

Dear Margaret,

I just finished your book and my first thought? Quote me on this: "This is the best book I have ever read." I wept openly as I finished the book and mostly because every sentence, every word — I lived those experiences and you have truly captured my era. I also cried as I did not want the book to end. Now I have to go back to the beginning. I received your book Saturday, read the first few pages, and very reluctantly had to put it down as the first of my seven color appointments had arrived and was waiting in my chair. As I always do, after I got this new high-school majorette client settled and into the routine of color foil highlights, I started to tell her about the book. I showed her the cover and let her read while I foiled her hair. Next client, my niece, same thing, then I had a whole crew of my high-school daughter's friends who all come in together to get their unique colors done. As I was foiling Adam's black and deep crimson foils, I asked Lindsey, whose color was processing, to read from your book. She started at the beginning with "January 1. It rained today. We were going to the movies." By the time she had gotten to the second page, all the kids were laughing and hanging on every word. Now I was doing color on Adam, 17; Lindsey, 15; Mallory, 16; Emily, 16; with their friend Nick, 17, watching. My youngest daughter Madeleine, 12, and her friends

Kaitlyn, 12, and Taylor, 11, were also in various stages of flopping on the floor or leaning against the walls while sitting on the floor. I had a shop full of teenagers and they weren't just listening to your book. They were having the time of their lives. By the time you got to saying the word "shitty crazy" they were hooked. I have never seen kids have so much fun hearing a book read aloud. Now I am crying as I write this. Because of your book, I have had the best two days of my life. I was able to time-warp back to when I was a teenager (just turned fifty and hot flashing) and I was able to be included in my teenagers' and their friends' lives. I so endorse this book that I am "shitty crazy" going to do just that.

Now, after we read, seriously, half your book out loud, one of my daughter's friends insisted on reading silently. We about killed her as she refused to read aloud and she kept laughing hilariously. Every time I read a book it changes me. I become a much better person. With your book, I feel as if I have been "born again" and inspired to reenergize my efforts on my mission to promote literacy

and get America reading.
Anxious to hear your reply,
Kathy L. Patrick
(known as "Murphy" in high school)
Founder of the
Pulpwood Queens Book Clubs

That message was just as I wrote it, off the top of my head. I meant every word and still do. Books empower us. I remember reading former president Bill Clinton's book, *My Life,* and being very impressed with all the references he made to books. Look where reading and education took this small-town Arkansas boy, regardless of your political beliefs.

Speaking of heads, I just happen to believe that by crowning ourselves Queens and Kings, we empower ourselves to rise to the occasion. And just tell me truly, who hasn't wanted to be crowned Queen or King?

> "Big things are expected of us, and nothing big ever came from being small."
> — *Bill Clinton, forty-second president of the United States*

Once a woman came in my shop and

oohed and ahhed over every piece of glittery jewelry or beaded bag she found. I was completely taken with her childlike glee. We got to talking and I learned that she and her husband were on a getaway anniversary weekend in historic Jefferson, the bed-and-breakfast capital of East Texas. As she handed me an armload of things she wanted to buy, she said, "I am a prison guard. I never get a chance to indulge my feminine side, and this weekend I'm making up for lost time."

Most of us don't work in a harsh environment like a prison, but we all need to dress up and feel like a Queen sometimes. As long as you keep your priorities straight, I am all for adding a little zing to our lives every now and again with clothes and jewelry.

When I was little, Bi-Rite, our local five-and-dime, sold a beauty-queen set that I wanted more than just about anything else in the world. It came in one of those flimsy cardboard boxes with a cellophane lid, the better to show off the treasures inside. I coveted the items the package presented: a "mink," two high-heeled sandals made of see-through plastic with stretchy elastic glittery silver straps to hold them on your feet, a princess's scepter, and a real silver plastic tiara.

> " 'There is no use trying,' said Alice. 'One can't believe the impossible things.' 'I dare say you haven't had much practice,' said the Queen. 'When I was your age, I always did it for a half an hour a day. Why, sometimes I've believed as many as six impossible things before breakfast.' "
> — *Lewis Carroll's* Alice's Adventures in Wonderland

The Queen for a Day set was out of my price range, which made me want it even more. My mother did not approve.

"But, Mommy, everybody has them," I whined. I begged. I pleaded. My mother didn't budge, pointing out dismissively that the fake mink looked like rat's fur. I would never admit that I thought so too. Hey, fur was fur and I wanted one. She informed me that the plastic sandals would break after one or two wearings and were probably too little for my big feet, and the tiara? That plastic piece of trash contained no jewels, not even fake ones. Okay, okay, so it was made from a mold with bumps and indentations meant to look like jewels. I knew those kind of tiaras never seemed to stay on your head for more than a few seconds at a time.

But come on, I was looking at the big picture. I could be a Queen!

A child's imagination is a powerful thing. Remember that this was the early sixties. Miss America reigned supreme, and Bert Parks's famous rendition of the pageant's theme song could be heard wafting through the school halls and on playgrounds, wherever groups of little girls played. Every little girl I ever knew of dreamed of wearing the crown herself one day.

At the height of its popularity, the televised Miss America pageant was as popular as the Oscars are today, and all over America girls and women could not wait to tune in for this once-a-year night of magic and glamour. My, how times have changed. Today not one of the major television stations even carries it, and I read recently that cable picked it up. It's not quite the same anymore.

To my mother, my two sisters, and me, the pageant was a special occasion. Every other night of the year, my mother had us in bed by eight-thirty sharp. Not the night of the Miss America pageant. We had to take our baths and get into our pajamas before the pageant started, so we could go immediately to bed when it ended. But until then, it was the Murphy girls' annual Miss

America–watching party.

My mother would make a big pitcher of iced tea and pop a big bowl of popcorn, which we would spread out on the coffee table in front of our nubby green divan. My mother always turned off the lights to make the living room feel more like a movie theater. We sat in a row in front of the TV, just like most of America did then with their TV trays, giddy with excitement. Finally, the commercials ended and Bert Parks started singing his famous song as a show-stopping razzle-dazzle display of the contestants paraded back and forth across the stage dressed in costumes that represented their home states.

You probably think that now I'm going to tell you we sat there and cheered for Miss Kansas. Well, to be honest, we always found our state's entry a little plain compared to, say, Miss Texas, so we weren't into showing state loyalty. We didn't exactly choose other favorites, either. No, it shames me to say it, but we Murphy women were irreverent. We made fun of all the girls.

Miss Kansas would stride down the runway in her bathing suit, smiling her best million-dollar smile, and my mother would say, just as catty as could be, "Girls, just look at her waist. That's what happens when

301

you don't cinch it in with a belt."

I would cringe at this barb as she gave me that raised-eyebrow look and pretend I had dropped a piece of popcorn until I felt her gaze go away. My sisters and I would do our best to figure out what was so bad about Miss Kansas's waist. It looked normal to us. As far as my mother was concerned, anyone who did not have a twenty-one-inch waist like hers had let herself go to hell in a handbasket. That's all there was to it.

In the spirit of female camaraderie — and not wanting to do anything that might get us sent off to bed — we didn't disagree. Instead, we joined in the fun.

"What's on her head, Mommy? A helmet?" one of us might cry out about some poor Southern girl's teased hairdo, collapsing in a fit of laughter at our own cleverness. My mother would smile approvingly.

> "Too bad all the people who know how to run the country are busy driving taxicabs and cutting hair."
> — George Burns, American comedian

And if we heard one more girl announce that all she wanted was "world peace," we would all just about throw up. No matter how much pleasure we took from pointing out those poor girls' imperfections, when I think about it now, it is obvious that it was jealousy on our parts, pure and simple. We were just too ignorant to realize it at the time. My mother's cattiness grew out of her own frustrations.

I know Mother watched that Miss America pageant thinking it should have been her up there, not Miss Kansas with the thick twenty-four-inch waist. The truth is that we would have traded places with any of those girls in a New York minute. We knew in our hearts that they were as beautiful and poised as we wanted to be.

Hence my intense desire for Bi-Rite's beauty-queen set. It was as close as I was ever going to get to feeling like a Miss America contestant, or so I believed at the time.

"Please can I get it, Mommy? Please?" My sisters and I would beg, in turn, whenever we wandered through Bi-Rite on our trips into town. However, no matter how much my sisters and I pleaded and begged, my mother was not having any of it. "No, girls," she would say. "That is an inferior-quality product."

While we were growing up, my mother bought almost all our clothes from the Sears

303

catalog. My grandmother Murphy made the rest, except for some tremendous sales from Zenishek's Department Store or the Frock and Bonnet fancy party dress and accessory shop. For anyone old enough to remember, the Sears catalog was a tome about as fat as a big-city phone book and just chock full of every kind of goody imaginable. It also made a heck of a booster seat.

> "Just the other day my assistant was on the line with Calvin Klein. I usually shop at Sears."
> — *Billy Bob Thornton*

Its seasonal appearance in the mail was a major event in many a household in the 1950s and 1960s. Sears had something for everyone in the family, including clothes, jewelry, candy, uniforms, major appliances, cameras, radios, toys, housewares, tools, plants, garden supplies, baby carriages, bassinets — you name it. If you needed it, more than likely you'd find it in the Sears catalog.

The Sears Kenmore brand washing machines and power tools were widely considered the best — well made, reasonably priced — and they came with a lifetime guarantee. Lots of folks *still* swear by Kenmore products, including me; I love to shop at Sears. Give me a Sears Craftsman tool any day. I just wish they made them with hot pink or leopard handles instead of that tacky red.

I spent hours thumbing through those pages, zeroing in on the children's clothing section. I would circle all the things I wanted with a blue ball-point Bic pen. Then my mother would decide which of my choices met her approval. Not all of them made the cut. More than likely my selections were chosen from the Sears sale circular. I never knew what I was actually going to get until the package arrived. I knew better than to express my disappointment. Once the clothes were delivered, my mother didn't have too much say in how I put it all together. I may have been a tomboy, but I liked to dress with an artistic flair.

In grade school I was to wear dresses, skirts, and jumpers, all of which fell properly just above the knee. You don't see too many young girls wearing jumpers nowadays. Blue jeans, T-shirts, and tennis shoes are what both sexes wear to school today. I cannot even imagine what our older teachers and principals would have thought about the kids' clothing now. I can only wonder as I look at my high-school-age daughter, in her black skinny tee, black pants with plaid black and pink suspenders hanging over her bum, black and pink Converses, cat collar with bell and tag that reads PSYCHO KITTY around her neck, and black fuzzy cat ears stuck in her Chi flat-ironed straight blond hair with hot pink streaks.

My favorite outfit in grade school was a hot pink velveteen jumper, which I always wore over a white Peter Pan–collared blouse. Despite the hot pink, the ensemble was appropriately demure, and it sailed under my mother's radar. I liked to pair hot pink and purple, and so I wore my hot pink jumper with purple knee socks. The real attraction — the element that turned this simple outfit into a flat-out fashion statement — was the knee-high see-through go-go boots I wore with it.

This was 1966. Thanks to the Beatles, English mod style was all the rage. In second-period social studies, I may have been listening to my teacher going on about the Boston Tea Party, but in my mind I was Nancy Sinatra singing, "These boots were made for walking."

On cold days I would complete my fashion statement by wearing my moss green plaid coat with the attached fringed scarf and a

> "The hair is real; it's just the head that is fake."
> — *Steve Allen, American comedian*

real leopard-skin beret with a black python headband that I pilfered from my mother's walk-in closet. I must have been quite a sight, as most of my classmates preferred plain, more subdued tones. I felt just like a princess when I wore that outfit. I knew I looked dee-vine! I reminded myself of Sandra Dee in that Gidget movie where she went to Rome. My mother, probably happy that I was finally showing a girlish interest in clothes, never complained.

After lunch, we played forty-five after forty-five record in the school gym and practiced all the new dances: the Pony, the Freddy, and the Jerk. My best friend, Debbie Teegardin, and I used to swing-dance and we just loved to do the Monkey. I was particularly fond of the record that played "Wipe Out!" and anything by the Beach Boys. We had to practice our dancing because Miss America always had those big dance numbers. Besides, Debbie and I always dreamed that when we grew up we would be featured on American Bandstand.

In those days we weren't allowed to wear pants to school. Finally, when we got to junior high, the school board decided that if the temperature reached ten degrees below zero, then we could wear pantsuits. That did not mean jeans or even slacks. That meant sweater-and-pants outfits, usually wool with fully matching dress pants that were lined and had side zippers.

When I was a sophomore in high school, my mother entered me in the Miss Eureka Pageant. I was never so horrified in my life, since only the really, really popular senior girls entered that competition. Nevertheless, I was going to be in it and all I really had to do was wear a nice dress. This was no formal pageant with bathing suits and evening gowns, thank God. All I had do was go to the courthouse for an interview with the judges.

Wearing my absolute favorite dress that year, a purple suede-ish empire-waisted, micromini dress with puffy sleeves, I arrived with my long blond-streaked hair recently set on hot rollers, wearing suntan pantyhose and my purple clunky shoes with oversized gold buckles. I thought I looked fairly decent. Now my kids would say, "Mom, you looked whack!" At the time I imagined I looked just like the model in the magazines advertising Prell shampoo, squeaky clean and like the girl next door.

The judges interviewed us and the only question I remember to this day is "Kathy, can you tell us your measurements, please?"

"Yes, they are 36-34-36."

The older brother of one of my best friends was a local insurance man and a judge. He started guffawing and laughing loudly. I was mortified beyond all comprehension. All the other male judges started laughing, and I just pretended it was funny too. Inside I was dying. I was smiling to beat the band, but what I really wanted to do was run from the room bawling my eyes out. I blanked out on the rest of that day. I do remember that the winner was Sue Kennedy, who was a senior, really pretty, and a nice girl. She was crowned Miss Eureka. I went home and wouldn't speak to my mother for weeks. That was another notch in my mother's super-skinny belt reminding me that I should have worn a belt when I was little.

Years later, as a bookseller/buyer, I attended a book-launch party in Beverly Hills for Berry Gordy, founder of Motown, for his book *To Be Loved: The Music, the Magic, the Memories of Motown*. Around midnight, who would walk in but Dick Clark. I and my fellow booksellers from Texas boldly walked up and introduced ourselves to Berry Gordy and Dick Clark. Besides Mr. Clark, we girls were the only white people at the club, but did we ever have fun. I was standing right in front of an orchestra that was playing earlier in the evening when Berry Gordy walked up to the microphone and announced that his good friend Smokey Robinson was going to sing for us. I stood right in front of Smokey Robinson for an hour as he sang, it seemed, right to me. Never in my wildest dreams would I ever believe that, coming from a small town in Kansas, I could have an evening like that. I never will forget meeting the man whose show we turned on every Saturday to hear the latest in music and see the latest dances. Now, what other profession can you have where you get up close and personal with the stars? Yes, books have opened some incredible opportunities for me. Not only have we book lovers crowned ourselves Queens, but we have begun to mingle with the stars!

So I guess it's no surprise that I chose hot pink and leopard print as the Pulpwood Queens' signature look. There is nothing more fun for girls than playing "dress up," and I have given license to my Pulpwood Queens to do just that. As adults we are always so aware, especially as middle age creeps upon us, that time is not always on our side. We don't wear the same size clothes we did when we were younger, but we still want to feel pretty and think we could catch someone's eye. Believe you me, the Pulpwood Queens, with their tiaras perched on their big hair, get plenty of looks and wolf whistles.

You really want to know what it feels like to be a Queen? Wear a tiara! When I do, I am always escorted to the front of the line and treated like a Queen!

As for diamond tiaras, I'm happy to say my desire for that beauty-queen tiara eventually moved beyond cheap dime-store imitations. The official Pulpwood Queens tiara is the real deal, the best rhinestone version money can buy.

In fact, I have upgraded my tiaras each year, one costing almost two hundred dollars wholesale. On my fiftieth birthday my husband bought me a darling vintage one from Rouge Beauty in the French Quarter in New Orleans. I love that tiara. It reminds me of something my favorite Queen, the famous country singer Dolly Parton, once said when an interviewer commented on an over-the-top rhinestone outfit she was wearing. Nonplussed by the subtle dig behind the question, she laughed her girlish giggle and answered with typical flare: "You have no idea how expensive it is to look this cheap."

As I've said, there was, at first, a little resistance on the part of some of the Pulpwood Queens to adopting our uniform, for lack of a better word. It taught me something valuable. Namely, that a lot of women think leopard print is sexy and would secretly love to wear it. They're just too embarrassed. "Kathy, I could never," they'd say. "It's like saying, 'Look at me, I'm a sexpot!' " Maybe that's why leopard print has always been considered a little on the trashy side. Not long ago, writer and official Pulpwood Queen Girlfriend Iris Rainer Dart sent me a photo of her and Cher on the set of the old *Sonny and Cher Comedy Hour,* the hit comedy music and review show from the 1970s, for which Iris was one of the writers. I remembered Cher's character "Laverne" dressed in a skintight leopard-print jumpsuit, stretch silver belt, dangling bobble earrings, black rhinestone cat-eye glasses, and a red, curled "big hair" hairdo, smacking gum just as loud as could be. I think it has something to do with the fact that leopard is the ultimate sexy fashion statement, to be found on femmes fatales and ladies of the night, and not fit for rock-solid, churchgoing wives and mothers.

But you'd be surprised how fast they came around. "Leopard skin?" Pulpwood Queen Auntie Bev whispered dangerously the first time she heard about it. "Oh, I don't know. After all, that's a little on the wild side, and what would people in church say? I'm a Baptist!"

> "My weakness is wearing too much leopard print."
> — *Jackie Collins*

She thought for a minute, then added with a wink, "Make that backdoor Baptist. Where's my leopard jacket?"

Now you have got to know Beverly, as all the above was said tongue in cheek. She is the one girlfriend I always grab when I have to be featured on television because, girls, she is Miss Kitty from the *Gunsmoke* television show come to life. Every year she

313

plays the jezebel in our infamous *Diamond Bessie Murder Trial* put on by the Excelsior House Players. She can do dance-hall girl better than anybody and can belt out a song that has gotten her chances to open for such country greats as Lee Greenwood.

Brazen, bold, not afraid to say what is on her book-reading mind, Beverly is the epitome of a real Texas woman. ONE HOT MAMA is what it says on her double-cab, deep-cherry-wine pickup truck that screams, "Girls, I have pimped my ride." She is absolutely "stop you in your tracks" gorgeous and also just happens to be my aunt. Funny, she is younger than I am, but she did marry my husband's mother's brother. Only in Texas!

Beverly is typical of the Pulpwood Queens membership. Here is a woman who realized that her child Brooke would be leaving the nest for college soon and instead of crying, she went back to college and got her real estate license. Yes, Auntie Bev, or Boom Boom, as she is known to her fans, is now the Official Realtor for the Pulpwood Queens. She can read our books and get us a good deal on a house, too!

Every year since I started the Pulpwood Queens I host a birthday bash for myself, and you will not see me putting "In lieu of gifts, blah, blah, blah . . ." on the invitations. I say bring it on. But that's just for fun. What I'm really saying is, "This is my day, and it is worth celebrating." We are taught our whole lives to say, "No, thank you." Or, "I couldn't accept such a generous gift."

> You can never wear too much leopard print; it's an instant icebreaker for making new friends. Wear a tiara and you will be treated like a Queen.

That it's better to give than to receive.

"Nonsense," I tell the Queens. My motto is, "It's better to give, but it's also really fun to get presents!"

Now, don't get me wrong, I love to give presents as much as I love to receive them. The handmade items are my favorites, and I still howl over the leopard-print camera Pam McGregor got me in Mexico. This year her bag of goodies included a rock — yes, a rock. I will cherish it forever. It is the thought, not the amount spent, that counts.

This year for my big five-o, I asked each of my Pulpwood Queens, authors, friends, and family to give me a hand mirror signed by them to hang in a row as a border around

> Life is too short not to have a *grand* sense of humor and adventure, so why not help a great cause while you are at it!

the walls in my shop. I told them they need not be expensive, just indicative of their personalities and their books. I wanted to have a reflection of my friends that I could gaze at while sitting at my desk. The mirrors come daily. I have to give author Loraine Despres credit for her originality. She sent me a mirrored license plate that read HOLLYWOOD, autographed by her.

As much as I love presents — the giving and the getting — I would rather spend our big money on projects that promote literacy. That is why the Pulpwood Queens are involved in community projects to promote literacy: Just like Miss America, we believe in giving back to our community. For three years running we have held the auspicious title of Most Spirited Team in our Annual Relay for Life, which is a twelve-hour (7 p.m. to 7 a.m.) Walk for the Cure sponsored by the American Cancer Society. This year we even raised the most money, over

Mandy's Adventure with the Queens!

My sister, LeTricia, a charter member of Kathy's book club, invited me to a MAW party (an acronym given us by Anne Dingus, a contributing writer for *Texas Monthly* that stands for Miss America–Watching, a party she holds with her girlfriends every year) in September 2001. Several times before, she mentioned her wonderful book club, shared great novels, and bragged on Kathy's enthusiasm, so I eagerly agreed to attend the Miss America–Watching party with her in Jefferson.

After agreeing to make this a family affair, my mother and grandmother voyaged to Goodwill to purchase secondhand gowns. Mom salvaged an iridescent frock channeling Charo and Pretty Prom Barbie, while my eighty-six-year-old grandmother chose a timeless little black dress. Tricia and I recycled bridesmaid dresses paired with leopard bras and boas. Clad in our best attire, we arrived at the Patrick Plantation and stuffed our faces like white trash at the Golden Corral, feasting on the three major food groups: chocolate, cheese, and wine.

Together with several other Queens, we criticized and complimented each lovely contestant. Whoever guessed the correct winner and runners-up was to be crowned Pulpwood Queen 2002! Well, yours truly won the crown with my mom as a very close runner-up!

Years later, after starting my own Pulpwood Queens chapter and making very valuable friendships, I can say that the Pulpwood Queens Book Club has been, and still is, much more than just promoting literacy and sharing great books over cheese and wine. I've made a lifetime of friends and I've contributed to the literary growth in my community. Most importantly, my goal is to impact lives around me with a positive Pulpwood Queens smile on my face and a tiara on my head!

— *Amanda, of the Pulpwood Queens of Rustin, Louisiana*

$6,000, thanks to our sponsor Lodi Drilling and the wonderful David McKnight and to Kathy Whitfield of Century 21. And let me tell you, we were on the track walking all night for the cure, but we had the most fun doing the walk with one another.

Crown yourself Queen and read the following:

Miss American Pie: A Diary of Love, Secrets, and Growing Up in the 1970s **by Margaret Sartor** The highlights of the author's actual diaries from ages thirteen to eighteen brought me and my Splinters teen chapter of the Pulpwood Queens to tears, laughing and crying. Teens *love* this book!

Savannah from Savannah **by Denise Hildreth** Savannah goes home to make her mark in a world where her mother reigned as Miss Georgia United States of America. It's a great mother/daughter read, and now there are two more Savannah books: *Savannah Comes Undone* and *Savannah by the Sea!*

The Sweet Potato Queens' Book of Love **by Jill Conner Browne** Jill Conner Browne rules as *the Queen* who does it her way and gets applauded for her efforts! Jill is funny, witty, and wise and knows how to party with her girlfriends better than anybody. We can all learn that it is fun to "play" even when we become grown-ups!

Because of my eccentric upbringing, I developed a certain taste for books that are a bit dark-humored or quirky. Here are some other outrageously royal reads:

Splendora by **Edward Swift** A splendid book about a young man who inherits the family home and moves back, only now as a woman and the town's librarian. Funny, real, and it would make the most incredible movie. Look out, *Tootsie!*

Paradise Screwed by **Carl Hiaasen** Check out the first sentence: "A true tale from the Miss Universe extravaganza: On Sunday, something called a 'Squirtmobile' was supposed to 'dispense' suntan lotion all over 10 of the most beautiful women in the world — this according to the official Miss Universe press apparatus."

This collection of over 200 columns picks up where his last collection of columns, *Kick Ass: Selected Columns of Carl Hiaasen,* left off.

Kick Me by **Paul Feig** My friend and Pulpwood Queen Carol Lancaster Lucky found this book kicked aside on the floor of the shipping and packing room at the trade publishing event Book Expo. We read the book together on our trip home, and almost got kicked off the plane because we were laughing so hard we were disturbing the other passengers!

***Down from the Dog Star* by Daniel Glover** Daniel Glover wrote this book as a journaling process as he recovered from a thirty-year heroin addiction. I laughed so hard when I first read it that I called him immediately to tell him, and invited him to come for a visit. When my coworkers asked what in the world a Southern gay writer was doing staying with us, my husband simply stated, "Old college roommate." That's why I married the man: we have the same sense of humor.

***Son of a Preacher Man: My Search for Grace in the Shadows* by Jay Bakker** Can you imagine being the son of Tammy Faye and Jim Bakker? Jay's parents' fall from grace came when he was only eleven years old, and this book will lift your faith and belief that God does have a purpose for each and every one of us.

***Lord Vishnu's Love Handles: A Spy Novel (Sort Of)* by Will Clarke** How do I explain a book about a man who thinks he's psychic, is facing financial ruin, and is on the brink of being insane, whose family hosts an intervention only to have him meet a Hindu holy man in rehab and decide he is going to save the world?

Electroboy: A Memoir of Mania by **Andy Behrman** Andy Behrman seems like a man who has it all together, running a multimillion-dollar publicity agency. Little did anyone know that he was battling for his life with manic depression and bipolar disorder that led him to excessive sex, drugs, and art forgery.

Midnight in the Garden of Good and Evil by **John Berendt** One of the most unusual nonfiction books I have ever read — and the book that made Savannah famous. But the most amazing thing is that this book was printed verbatim, exactly as John Berendt wrote the original manuscript. No editing whatsoever! He is a genius!

CHAPTER 11
I CAN CHEW GUM
AND READ AT THE
SAME TIME!

"It takes a smart brunette to play a dumb blonde."
— *Marilyn Monroe*

Back in 1976, when I was a student at Crum's Beauty College, Mrs. Crum asked me to speak to the Manhattan Rotary Club about what it was like to be a beauty-school student and what was involved in becoming a salon professional. When she approached me about it she announced, "Ms. Murphy, I have scheduled you to speak to the Manhattan Rotary Club and you will be excused from classes to attend this function."

I turned around to see who might be standing behind me, because I thought she had to be talking to someone else. She continued with her jet-black eyebrows arched and hands on her hips. "You are to speak for fifteen to twenty minutes on what it is like to be a student at my beauty school.

They will be serving you lunch, and I am sure you are more than up to the task of representing our school in a positive and professional manner."

There was no arguing with Mrs. Crum. When she spoke, you listened. You did her bidding or else. I certainly never wanted to find out what the "else" was as I watched her point her blood-red fingernails at me while she filled in the directions to the Manhattan Rotary meeting.

Public speaking was not my forte. In fact, I practically flunked speech in high school because, basically, I just couldn't do it. I could talk your ear off in a small group of friends or one-on-one, but standing in front of a sea of faces, especially boys, caused beads of perspiration to appear on my upper lip. My face would flush bright pink and my mind would go completely blank. "Uh . . . um . . . uh," I would stutter, wishing I could become exactly like the Invisible Man and disappear. My mind would become a blank screen. After a tortured few minutes, which seemed like hours to me, Mr. Peacock, my high-school speech and drama teacher, would dismiss me to my seat in total disgust.

Today I speak in public all the time, as I am often asked to talk to business and civic groups, and many people tell me they admire the ease with which I do it.

"You can thank Mrs. Crum," I say. Ignoring the perplexed look on their faces, I go on to explain that I owe any success I have in this area to Mrs. Crum, who looked past my shyness to see a natural-born speaker hiding in the wings. Mrs. Crum gave me advice that I now know is the secret to public speaking — and to life, if you really think about it. She told me just to get up and talk about my experiences at the school. I was to talk about what I knew.

Even though I felt as nervous as a cat in a dog pound, I waited to be called to the podium. As the president introduced me, I stood and approached the microphone. One of the instructors from the beauty school had come with me. She just smiled, nodding at me to go ahead. When the polite applause died down, I opened up my mouth and the words started pouring out. Before I knew it, I was in the zone — relaxed, connecting to my audience, and actually enjoying the experience. This was my amazing discovery: If you talk about what you know and you really believe in what you are saying, giving a speech is as easy as pie. I talked about what I was learning at Crum's Beauty College. I talked about what we did as students on an average day and what our goals were to complete the course. I talked about what I planned to do after graduation.

Back then, the Rotary Club was all men. For the first time, as I looked out into the sea of their faces, I felt like an adult. They listened intently, then gave me a standing ovation at the end of my speech. Many members came up to me after the presentation and told me what a wonderful job I had done as they extended their arms to shake my hand. I was shocked and surprised, because this was not what I had been expecting. I left the meeting feeling on top of the world.

"All of us are born with a set of instinctive fears — of falling, of the dark, of lobsters, of falling on lobsters in the dark, or speaking before a Rotary Club, and of the words 'Some Assembly Required.' "
— *Dave Barry, American writer and humorist, best known for his weekly newspaper column*

When I got back to the beauty school, Mrs. Crum called me into her office.

"Ms. Murphy, I have had a full report from the Manhattan Rotary Club, and you did an excellent job representing my school. I am sure we will have many more opportunities for you to represent us. Congratulations, and now you may go back to your class. You are excused. Please shut the door on your way out."

I practically skipped out the door. Never had I felt so confident! As I went back to class, I kept replaying the scene in my mind. Mrs. Crum's words continued to play in my head as a positive mantra. When I had given speeches before, I'd had to talk about a subject that didn't fully engage me and had had to memorize the speech because I couldn't get it to stay in my head any other way. I was filled with anxiety. Isn't it interesting that, whether you are talking about choosing clothes or a hair color, or a job or a hobby, it all comes back to being true to who you are?

Mrs. Crum believed in me in spite of myself and reminded me of the profound truth that cannot be said too often: When someone believes in you, a world of possibility opens up before you. Mrs. Crum encouraged me to do things that, left to my own insecure devices, I would never have tried on my own. It was just like the time she had me enter the state competition for hairstyling and told me that it did not matter whether I won or lost; I'd learn from the experience. She also told me I was the best representative from the school. I had never been told I was the best at anything before.

When I spoke to the men of the Manhattan Rotary Club, something very extraordinary happened. At least it was astonishing to me at the time. Back then, being tall and blond, all legs and boobs, I was used to being treated like a dumb blonde by boys and some of the teachers too. The Rotarians listened to me. They asked questions and treated me as an adult, with respect. My confidence rose to the occasion.

Years later, when the media discovered that Beauty and the Book was a hair salon/bookstore, one reporter from New York called for a telephone interview and had the nerve to ask questions like "Are you chewing gum right now? Exactly how big do you make your clients' hair? Wait, I can't hear you for all the noise from those hair dryers."

Another Southern woman reporter asked me, "If I come to your shop, will you backcomb my hair *really* big like you wear your big hair over there in Texas?"

The worst question yet was from yet

another good ole boy, who asked, "Hey, got any good dumb-blonde jokes?"

I used to wince at these questions, but I tried to answer them politely nonetheless. I would explain, "Actually, I hardly ever do 'big hair' on my clients unless they want it for a special occasion, like portraying Julia Roberts in *Steel Magnolias* for Halloween."

> "My real hair color is kind of dark blond. Now I just have mood hair."
> — *Julia Roberts*

I remember once all these girls came in to get their hair done for a wedding. I was to do the bride and three or four of her bridesmaids. The bride wanted a more traditional classic updo with her veil. The bridesmaids just had set their hearts on having big hairdos, so she wished them well. In all my many years of doing weddings I have never seen college-age girls have that much fun. After I had whipped up the frothy concoction of big barrel curls, one bridesmaid cried ecstatically, "You can see through the curls all the way through to the back. Oh my Gawd, I love my *hair!*"

The other girls were on the floor in hysterics, practically crying they were laughing so hard. I can honestly say I never had more fun than doing that wedding. Those girls were celebrating the wedding of their girlfriend the right way. They were having some Big Time Fun with their Big Time Hair.

In fact, Marla Keith, who works with me in the shop, and I spend an enormous amount of time, energy, and money keeping ourselves up-to-date on the latest in industry developments. I won't chew gum because I tend to pop it, and even I hate that sound. As for dumb-blonde jokes? Why would they even ask such a question of someone they assumed *was* a dumb blonde? I would always say, "I am not dumb. I have spent an enormous amount of time taking colleges courses with a double major in art and geology, and I continue to educate myself every day. In fact, I love learning. I am a reader."

> I may be a blond small-town hairdresser, but I can still read and chew gum at the same time.

Eventually, I saw that I was being defensive and so I came to my senses. Hey, I am not on trial here, and who made these yahoos judge and jury anyway? This isn't even about me; it's about them. These were folks who bought into just about every stereotype our culture has to offer, especially that blondes are dumb and hairdressers are just bimbos. Ergo, a blond hairdresser who promotes literacy is an oxymoron! Let's not take her seriously.

Those reporters' questions only showed their "in the box" thinking. But I knew better than to blow those reporters off. So I started quoting my personal hero, the great Dolly Parton.

This is what Dolly had to say when asked what she thought of dumb-blonde jokes: "They don't bother me a bit. Because, first of all, I'm not stupid, and, second of all, I'm not blond!"

> "Find out who you are and do it on purpose."
> — Dolly Parton

You go, Dolly. I admire the woman so much. With her platinum-blond big hair and other outsized assets, her rhinestone clothing, long painted nails, and colorful makeup, she oozes confidence and doesn't really worry about what anyone else has to say about it.

I was inspired to be a Pulpwood Queen after reading *The Glass Castle* by Jeannette Walls. I never took the Pulpwood Queens seriously, and often laughed and even poked fun at their antics and appearance. However, as a media guest at the 2006 Girlfriend Weekend press conference, I was in awe when I saw MSNBC reporter Walls walking up the sidewalk at Kathy's bookstore/beauty shop, Beauty and the Book. To realize that someone I admired as much as Walls would be a part of an event such as this changed my thinking and forced me to turn my feelings around and realize that the work of Kathy Patrick was worthwhile and not in vain. After that weekend I started a Pulpwood Queens chapter in my hometown, and I have been wearing my tiara and reading ever since.

— *Phyllis, of the Pulpwood Queens of Marshall, Texas*

Most people know about Dolly Parton's physical attributes, or that she is a living country legend for her singing and songwriting skills, but what they may not know is that Dolly is a tireless crusader for literacy. Beginning with putting down her own hard-earned cash, she has provided books for children in Pigeon Fork, Tennessee, from the time they were born until they entered school. Knowing that education and literacy were important, she went a step farther by founding Dolly Parton's Imagination Library at www.imaginationlibrary .com. It is my dream to raise enough money to start that program here, in Marion County, Texas, and then to inspire my Pulpwood Queens to champion the cause in their own neck of the woods.

Dolly promoting literacy? You bet — because there is a lot more to Dolly than what you see. I love it that she's breaking that blond stereotype. She is one smart woman, and one I admire very much.

> "I'm not going to limit myself just because people won't accept the fact that I can do something else."
> — *Dolly Parton*

In a 1992 interview for a VH1 special, she explained:

There was a woman in our hometown, and everybody said that she was trash, you know, the town tramp. But I thought she was beautiful 'cause she had this yellow hair, and she had bright red lipstick, and she wore makeup and tight clothes and bright colors and red fingernails and high-heel shoes. And I thought, That's how I want to look! I thought she was beautiful. Nothing was ever bright enough for me; I mean, I love loud colors. So my look came from a very serious place. It was a country girl's idea of what glamour really was. The more people tried to get me to change my look, the more I realized people were looking at me. I thought that if I can hold their attention long enough for them to see that I am a talented person and that I am sincere about my work and that there is a brain underneath all of this hair and there is a heart underneath these boobs, you know, and if people can really see beyond that, then I can get past that.

End of story. I think that is about the most profound thing anybody has ever said regarding being who you are and bursting through silly stereotypes. And did you know that besides being so gosh-darn gorgeous and talented, Dolly Parton is also one of the richest businesswomen in the United States? Some youngsters grow up dreaming of becoming president of the United States. When I grow up, I want to be just like Dolly.

I just love that the Pulpwood Queens are breaking down stereotypes. Of course, like most of us, I bought into those stereotypes at one time too. In fact, I think that stereotypes are so prevalent in our culture, especially the ones about how females ought to look and act, that most of us start believing them when we are too young to know any better.

When I first started working in the book business, I tried to be as far from the image of "dumb blonde" as possible. I colored my naturally dark blond hair black cherry and cut it off into a Louise Brooks bob. Although I was nearsighted, I started wearing the black Buddy Holly distance glasses that I usually only wore while driving, trying to look smarter and probably wrecking my vision in the process. The girl who had refused to wear a belt or anything constricting her whole life even went so far as to buy a minimizer bra so she wouldn't look too busty!

Have you heard the saying "When you try to please everybody, you please no one"? When we limit ourselves by trying to be what we think people expect us to be, all we accomplish is stunting our potential. Trying to conform to someone else's image of what they think is correct — that's thinking inside the box. I was inside that box myself — with the flaps shut and taped. I had tried to make myself over into an image of what I thought would be more acceptable in the book business. So what if people just make plain wrongheaded assumptions about who you are?

Even though I was a constant reader and had been reading a book a day since childhood, I was afraid people wouldn't take me seriously as a bookseller if I didn't look the part. I took to wearing black because, as anyone in the book business will tell you, black is the color of choice for New York City publishers.

These were also my child-rearing years, and, after giving birth to two big, healthy Texas girls, weighing almost ten and nine pounds, respectively, I embraced the more matronly look. My new favorite color black had a practical benefit: black always makes you look slimmer.

Still, the weight gain and forgetting to take care of myself happened gradually and without my willing it; it was a natural reaction to the constant, steady demands of raising young kids.

I basically adopted a new look because I

didn't think my old one was good enough for my new situation. I also studied the dictionary at night to improve my vocabulary. I was trying to reinvent myself and, except for the dictionary part, the whole look just wasn't the real me.

> It doesn't matter what size you are; stay active. Get up out of the chair after you finish a good book and go for a power walk. Your body and mind will thank you for it.

I was so busy with my job and the girls that I barely noticed that I was looking a little drab. That is, until I went out with Lainie and Madeleine and people started telling me just how cute my *grandchildren* were. Grandchildren! Jiminy Christmas! Suddenly I had this image of Madeleine graduating from high school, taking me shopping to buy Depends and moving me into the Cypress Valley senior assisted living facility here in Jefferson. At age *forty?*

I may technically have been old enough to be a grandmother, but I certainly didn't feel that old. That was my wake-up call. Back came my blond tresses — okay, enhanced a little this time around with highlights to help diminish the rapidly appearing crow's-feet forming around my eyes. I'm happy to say that they are there from all my years of laughing too hard. So if anybody comes at me suggesting Botox to alleviate my crow's-feet, I'll knock their block off. I earned every one of my laugh lines.

Women have been sold a bill of goods about wrinkles and aging. It's not just men who look better as they age. Just look at Annette Bening, a Kansas girl with the character and spunk to say, "I'm damn sexy with a few wrinkles." More Hollywood actresses should follow her lead. Women of America, unite! Be proud of your wrinkles — they are the map of the journey of your lives. Just Say No — to Botox!

So I got a stylish new haircut and went out and bought a bunch of new clothes in jewel tones. Those clothes made me feel alive. They were fun, exciting, and definitely not grandmotherly. Suddenly there was a new bounce in my step. I was liberated! I decided to lose weight.

I went to a doctor, who put me on a healthy new diet regimen. After my last child was born, I was diagnosed with gestational diabetes. I would always have to watch my weight and what I ate. Soon, I let the diabetic diet slide and began putting the weight back on. I started having horrible attacks that would leave me bent over, doubled up in pain. Back to the doctor, who found I had gallstones. I ended up having my gallbladder removed. Then I decided it was time to get really serious about getting healthy, and I went to a doctor who specialized in weight and eating disorders. She informed me I was the perfect candidate for a stroke and, effective immediately, put me back on a diabetic dict. I lost thirty pounds the first month with her carefully monitoring my weight loss at once- or twice-weekly appointments. I then noticed that everything was heading south — and fast. My body was soft, and my doctor suggested I start weight training and cardio to go with it. Had I ever heard of the House of Pain? I jumped at the name and said, "Are you serious? I am allergic to pain. Seriously, the nurse and my daddy had to sit on me when I had to have a shot at the doctor's. I don't do well with pain."

My smart doctor ignored my protests. "I am serious about the gym, Kathy. This is a hard-core gym that has one of the best trainers in the area and works with a lot of the patients at our local hospitals."

As she stood up to say good-bye, she instructed me, "Go to the House of Pain immediately, right today. Take your children with you, before you chicken out. Durrell Bowie is the owner and trainer, and he will get you all set up in a program that will fit you and your body type." So the kids and I drove to the House of Pain.

> "Reading is to the mind what exercise is to the body."
> — Joseph Addison, English essayist, poet, dramatist, and statesman

She wasn't joking — the House of Pain was a hard-core gym. I pulled tentatively into the parking area in front of a metal building that certainly was nothing to look at. In fact, it looked like it could have been a machine shop or a garage. As I parked I looked up at two of the largest men I have ever seen in my life. I was thoroughly intimidated by these men. As I stepped out of my car they looked at me as I asked, "Would either of you know where I could find Durrell Bowie?"

One of the men started walking to my van

as the girls got out. Though it was winter, both of the men had on skintight tank tops and form-fitting workout pants. Muscles bulging, they were obviously seriously into bodybuilding.

The smaller of the two — and that is seriously a misnomer, as he was huge — was fast approaching my van. "That would be me," he said as he reached out with his paw of a hand to shake mine. "Did Dr. Rasheed send you to me?"

Before I could squeak out a "yes," he had my arm and was leading me into the looming metal building with Lainie and Madeleine following us, whispering and snickering.

As I entered the gym, my eyes began to adjust to the dimmer light. The building was full of heavy metal gym equipment. Mirrors were placed randomly on the walls. I could see men pressing weights and middle-aged women running on treadmills; the gym seemed on the up-and-up, full of testosterone and macho-macho men.

My trainer introduced himself to my girls, shook their hands, then showed them where to stand to the side. They did exactly what he said.

"Here, get up on this machine," Mr. Bowie said as he proceeded to guide me up on a padded seat that I had to straddle in my dress clothes. I tried to object with "I am really not dressed."

"Wrap your arms around this padded bar and pull."

I wrapped my arms around the bar and pulled.

"Do ten of those pulls."

I did ten of those pulls. The girls were giggling.

"How was that? Too hard? Come over here with me. Get on this bike." I didn't even try to argue. I got on the bike. "Pedal."

I pedaled until he told me to stop. "You think you can do that?"

"Why sure, it really didn't seem that hard —"

He cut me off as he led me back to the door. "Be here Monday. What time is good for you?"

I told him mornings would be best as he led me to the van, opened the door, and let me in. The girls were smiling like hyenas and I knew I was going to get a ribbing once the doors were shut.

"See you Monday. Kathy, is it?"

I nodded yes and he was back off to the gym. I sat there in amazement.

"Wow, was that cool or what, girls?" I said as I started the car and looked behind me to back out of the parking space. The girls were howling with laughter.

> Everything you think, do, or say is wrapped around your state of mind. Tell yourself "Life is good, life is good"; the next thing you know, life *is* good! When you think positive and act positive, good things will come into your life.

"Momma, you were so funny."

My head was spinning from all that had happened in what — the last ten, maybe fifteen minutes. I thought if it was going to be that easy, why not? How hard could it be?

As I left the House of Pain that day, I felt better than when I walked in. After a few weeks, I was shedding pounds and getting pretty buff (for me), and started feeling on top of the world. I felt good about myself, and I noticed a change in the way people reacted to me.

One day at Piggly Wiggly, I was pushing a shopping cart loaded to the gills with the week's groceries through the automatic doors. A man who had just finished loading up his car had started the engine, and he was proceeding to back out when he saw me. He pulled back in, stopped his car, jumped right out, and rushed over to give me a hand. And I let him! I may have grown up in the era of women's lib and equal rights, but dang it, when you've got to juggle a full grocery cart, your pocket purse as big as a suitcase, and your whining kids, God bless their chauvinist hearts!

I am not athletic, but now I have endurance and I am strong. I love athletic activities like swimming, hiking, backpacking, canoeing, roller skating, snow skiing, and even weight training. Some people are shocked when they hear I do those things and say, "What — and break a nail?"

Errr. I can't do anything about them, but I can do something about me. The Pulpwood Queens leave our diets at the door, but we have the power to go power walking at 4:30 a.m. Challenge those assumptions!

> Don't be afraid to be yourself. God made you just the way you are.

Several years ago at the Marion County Fair, International Paper sponsored a timber competition. I sauntered up to the booth where several men in IP-logo shirts and jeans were getting the paperwork ready for the event.

"Where can I sign up?"

They looked up, surprised. I knew several of them. They decided to play along with what they thought was a joke. There must have been about twenty to thirty men gathered around looking at the logs up on sawhorses and checking out the handsaws and chain saws. They were all dressed in traditional East Texas attire: blue jeans or camouflage with "gimme" caps, the ball caps that are handed out as company advertising.

"I would like to enter your woodlands competition."

The men froze. A woman wanting to enter the timber competition? They all looked at each other, wondering what to do. One of the head honchos from the paper mill bravely spoke up: "You want to enter the competition?"

I could tell that the men were hanging on his every word.

"Yes I do, Roger. Exactly where do I sign up to enter?"

Several well-meaning and brawny men sputtered, "But this event is for guys only."

"Well, boys," I said, staring them down, "I would like to compete and surely you don't mean to tell me that just because I am a woman I can't?"

"Have you ever run a chain saw before?"

"Nope, I can't say that I have, but I cut hair. How hard could it be? A chain saw just looks like a bigger cutting tool to me." Batting my short, scrawny lashes, I mustered all of my femininity à la Marilyn Monroe. "If you big strong, husky guys could teach me, I think I could give it a try."

All the men started laughing and choking on their Dr Peppers as I took the clipboard that was handed to me to sign up. Roger, obviously in charge, walked up to me and told me just what to do. I was to put on safety goggles and strap on protective leg gear, kind of like chaps that cowboys wear. When they announced, "On your mark, get set, ready, go!" I was to pull the start on the chain saw and proceed to make a one-inch cut up the log and then another one-inch cut back down through the log. He told me to be careful — he didn't want me to cut my leg off. I assured him that I wanted to keep my leg. That wasn't even going to be an option.

The people who had the fastest time from the start of the judge's stopwatch to when they finally turned off the chain saw would win first, second, and third places. The men were to go first.

I watched them as they were timed cutting the logs. It really didn't look that hard to me. I certainly could cut off an inch of hair, so why wouldn't I be able to cut an inch off a log? The men were watching me and I think they were as nervous as I was. I was swimming in uncharted waters, and these guys weren't even sure I knew how to dog-paddle.

I had rounded up a couple more women to enter the competition. They let a friend of mine go first. She did pretty well, and I noticed she even beat some of the men's times. I snickered. Then it was my turn.

My husband had heard that I had signed up for the competition, and he joined the growing crowd of men. As two men began buckling the straps on the inside of my legs, my husband hollered, "Hey, that's my wife there — just watch what you're doing!"

All the guys started punching each other on the arms and backs and laughing. The adrenaline was starting to pump as I looked up. It seemed as if everybody from the Marion County Fair had come over to see what all the commotion was. We had, I think, half the fairgoing attendees crowded around the timber competition area. A few men were asking onlookers to step back and make some room. Roger motioned for me to step up to the log.

It was my turn, and I pulled the safety goggles over my face.

"Are you ready?" the men asked.

"Ready," I answered as I widened my stance to get a firm, steady bearing.

When they gave me the signal, I started the chain saw and began my first one-inch cut down the log. Little chips were flying, and it seemed to be taking me forever as I applied a bit more pressure. I was focused and now not a bit nervous. I was just doing the task I had set out to do. The first slice dropped and I began the one-inch cut up back through the end of the log.

Of course the chain saw was much heavier than a pair of scissors, but after I felt around to see how much pressure I could give the saw without losing control, it was easy. The second slice hit the ground and I turned off the saw. The judge looked at the time and then announced it. I had beaten the one woman and a good bunch of the men. Several more men then went on to compete as everyone watched to see who was going to have the best time. When the last man finished, several men and the timekeeper went off to the side with the clipboards to figure out who had won. The men's division was announced first, and then they announced I had won first place in the women's division. I was embarrassed as everybody started patting me on the back while the audience of men and women whooped and hollered.

> "Any woman who thinks she can win a tree-cutting contest because she knows how to cut hair is someone you need to get to know."
> — *Kathy Hepinstall, author of* The House of Gentle Men, The Absence of Nectar, *and* Prince of Lost Places, *all Pulpwood Queens Book Club Selections*

Say hello to the Marion County Chain Saw Champion — of the *newly formed* women's division. I didn't give a flip about winning. I can tell you, though, I was as proud as a peacock that I had done my little part to break through the gender barrier. And I don't mind saying that I have competed ever since, only to have the tar whooped out of me by other women who have entered the competition, mostly gals who worked for International Paper.

Who says a woman can't cut wood? Are

we physically as strong? Generally not, but I do believe that once a woman sets her mind to something, there is *no* stopping her.

Think outside the box. Throw away all your preconceived perceptions of what is status quo. Break through the stereotype and do something you have always wanted to do. There is nothing more powerful in the world than knowing that you tried. Forget those false assumptions and you just might find yourself setting a new trend!

Below are books sure to inspire others. I know they inspired me.

***Defying Gravity: A Celebration of Late-Blooming Women* by Prill Boyle** This is a collection of fascinating stories about women who found their true purpose and calling later in life — including the author, who actually became a writer in midlife.

***Delta Style: Eve Wasn't a Size 6 and Neither Am I* by Delta Burke** Who doesn't love Delta Burke, former Miss America and of *Designing Women* fame? She is one of my favorite actresses, and this poignant and telling book is inspiring to all women.

***Hero Mama* by Karen Spears Zacharias** Karen's father was killed in Vietnam, and her mother was left without a home and with a slew of kids and only an eighth-grade education. Karen tells her story in a non-judgmental way and now has made it her life's mission to help other soldiers' families.

***Live Like You Were Dying* by Michael Morris** I adopted this author as my brother because, I swear, he is the nicest man you will ever meet. This book could be the Pulpwood Queens' theme book, and the story makes you think about what is really important in your life.

***Wish It, Dream It, Do It: Turn the Life You're Living into the Life You Want* by Leslie Levine** Inspirational speaker and author Leslie Levine is a little dynamo who will not only inspire you, but reading her books will also make you a better person.

Here are some wonderful books about women who have often been stereotyped. I have found great inspiration in their beauty, art, and talent. Besides their special gifts, they also just happen to be women who are or once were blondes.

***Dolly: My Life and Other Unfinished Business* by Dolly Parton** This is my personal hero's story of her life of adversity and of determination to do it her way or no way. You have to give this woman credit for being a country music living legend and also for starting her Imagination Library not-for-profit foundation, providing books to babies and pre-school-age children.

***The Many Lives of Marilyn Monroe* by Sarah Churchwell** Since my mother was always fascinated with Marilyn Monroe, I have found that I am too. This book reveals Marilyn as truthfully and fully as I have ever seen.

***When I'm Bad, I'm Better: Mae West, Sex, and American Entertainment* by Marybeth Hamilton** Mae West was another blonde I adored as a child and still do today. She was ahead of her time, but growing up in show business will do that to you.

***Cybill Disobedience* by Cybill Shepherd with Aimee Lee Ball** Cybill Shepherd was a teen beauty queen with a brain and a mind. I found her story enthralling, and the tell-all part was pretty racy, too. I mean, she slept with Elvis.

***Elle Woods: Blonde at Heart,* with character created by Amanda Brown, story by Natalie Standiford** I have to admit I love Reese Witherspoon and saw *Legally Blonde* first and read the book second. I have always loved stories about blondes who get their day in the sun!

***The Million Dollar Mermaid* by Es-**

ther Williams Being a swimmer, I adored Esther Williams and her high dives and synchronized swimming in all those movies. Her real life is no less fascinating, and the part about Stewart Granger really made me sit up in my chair and pay attention. I always knew that Johnny Weissmuller was a lech.

Blonde Like Me: The Roots of the Blonde Myth in Our Culture **by Natalia Ilyin** This is a nonfiction book in which the author divides blondes into categories: the California Sun Blonde, the Trophy Blonde, the Moon Blonde . . . and helped me to understand why the platinum-blond Barbies are the biggest sellers.

Revolution from Within: A Book of Self-Esteem **by Gloria Steinem** A frank book on self-worth from the woman who has always fascinated me since she went undercover as a Playboy Bunny. Hey, she may have been our all-time favorite women's libber, but she was hot as that Bunny!

CHAPTER 12
MY OTHER BOOK
FAMILY

"Friends cross all barriers of race, creed, age, gender, and country to connect with the heart and spirit, which have no walls. Sometimes they don't even know it when they say the right words at just the right time. Sometimes friends *feel like* family. Sometimes they *are* family."
— *Donna Fargo*

I have a wealth of friends, and so do my daughters. I have friends with whom I work, friends who are my clients, friends who are Pulpwood Queens, friends who are authors, friends from church, childhood friends, friends from college, friends from Rotary. I have a lot of friends. But I have one best friend who is special. She is special because she's not involved with my work or my book club, as most of my closest friends are. We aren't friends from my school days or from college. She's special because of how we became friends: through our daughters. Her name is Mary.

My daughter Madeleine and Mary's daughter Kaitlyn are best friends. They have been since they were babies at Country Day School, since both their moms were working moms. When I met Kaitlyn's mom, Mary, it was probably at one of our girls' first birthday parties. As time went by the girls became inseparable, and Mary and I in turn got to know each other very well. What makes our friendship unique is that my friend Mary is exactly twenty years younger than I am — the same difference in age as between my mother, Mary (funny, they share the same name), and me. My friend Mary could be my daughter.

When I went to my twentieth reunion of Eureka High School's Class of 1974, I couldn't believe my classmates had children in college. My daughters were seven and four months old. Shoot, one of my classmates, Peggy Knudson, told me she had grandchildren older than my baby, Madeleine. I had the youngest child in my class. I was an older momma.

Recently, at my thirtieth class reunion, my classmates' children were graduating from college and most were married with children of their own. I often wondered what it must be like to have a child who is in her twenties, and then approaching her thirties. Having Mary as my friend, you would think I'd understand what it would be like to have a daughter her age, and in some ways I do. Yet as much as I would like to say I am like a mother to Mary, she is often more like a mother to me.

> "There is no friend like an old friend
> Who has shared our morning days.
> No greeting like his welcome,
> No homage like his praise."
> — *Oliver Wendell Holmes, Jr.*

Mary is a take-charge kind of girl. I admire her tenacity and spirit more than she will ever know. When I am with Mary, she is the one who rules the roost. I thrive on breaking stereotypes, and my friendship with Mary does exactly that. We don't have your typical soccer-mom friendship, and we actually don't have much in common. She's a night owl and I'm a morning person. She likes to organize and I like to fly by the seat of my skirt. I like to think we complement each other, like salt and pepper; we are both, in our own ways, the spice of life. I love to be with Mary because I just like hearing her talk, especially to my husband,

Jay. When Mary comes over, Jay — who is normally glued to the computer — comes bounding down the steps, and I just sit back and watch those two spar over everything from his not helping around the house to computer issues. They are both the funniest things I have ever seen. They are alike in their temperament and in their passion for anything involving technology, which is something I don't give a flip about. Mostly, I just like to hear Mary, to admire her confidence and her work ethic, and if I were to have a daughter twenty years my junior, I would wish that she were Mary.

Mary's life has never been an easy one. Mary's mother was Korean and married an American soldier, Mary's father. The marriage didn't work and ended in divorce. Mary and her mother moved to Texas to start a new beginning, leaving Mary's brother behind to be raised by his father in Oklahoma.

> "Adversity is the diamond dust Heaven polishes its jewels with."
> — Thomas Carlyle, Scottish historian and essayist

Wanting to provide the best kind of life

for her daughter, Mary's mother worked two jobs to make ends meet. She encouraged education for Mary, and Mary then had dreams of becoming a nurse. But just a few days before Mary graduated from high school, her mother was tragically killed in a car wreck in Longview, Texas. I had heard the story, but after years of knowing her I just came out one day and asked her what happened. She told me everything, and I was left spent and amazed that Mary had survived that tragedy so successfully.

Once I heard all the details I admired her more, because I do not think I would have had the strength Mary had to continue on and fulfill her dream of becoming a licensed vocational nurse. I couldn't believe such a tragic thing could happen to such a wonderful young woman on the cusp of finding her own place in the world. Mary persevered, put herself through nursing school, and had her daughter, Kaitlyn. Married to her high-school sweetheart, she continued to work in nursing and eventually had her son, Brent.

Our daughters became best friends when they were both enrolled at the local day care. They were inseparable from the time they learned to walk. At first Mary and I were like most new mothers, sharing child-care stories and ideas for birthday parties and planning playdates. But because Mary was young enough to be my daughter, I have grown to love her exactly that way.

Mary and her family are a part of my family. We celebrate birthdays and holidays together. I began to share my love of reading with Kaitlyn and Mary, and anytime an opportunity for a gift arrived, I gave them books for read-aloud times. Mary told me that when Kaitlyn started school, she instantly became the top reader in the class. I applauded her efforts and have never felt so proud. I am a firm believer, regardless of your political views, in the message Senator Hillary Clinton put forth years ago in her book *It Takes a Village:* that all of us are needed to raise a child. I have taken that message to heart, and though I have never actually put this into words to Mary and her family, they are a part of my village.

We recently took our girls to a Mardi Gras parade in Shreveport. Even though I was driving, Mary was the one packing the car with blankets, coats, and portable chairs for the event. She was the one determining our route as I drove with the map, making the calls for directions. When we arrived, she made sure everyone was bundled up properly, and I wouldn't have been surprised if she had personally tucked a blanket around my legs. I could argue with her and be the one who takes charge, but I honestly feel that she enjoys being in charge, so I let her. Sometimes I wonder what I actually bring to our relationship. Sure, I do her and Kaitlyn's hair. I love to give the kids books and have been known to read to them at the drop of a hat.

Before Brent was born, Mary told me how she wanted to do up Brent's nursery, gathering things the girls had outgrown in their own rooms, like a white rocker and a yellow moon lamp. Mary and I painted the nursery a sunny yellow. Then we hung a Noah's ark wallpaper border around the room. She painted the chest of drawers and replaced the drawer pulls with moon and star pulls. As we stood back to admire the room, I couldn't help thinking how much I'd enjoyed helping her put the nursery together.

When Brent was born, I remembered what it had been like for my oldest suddenly not to be the only child anymore, thrust into the world of sharing: sharing a room, sharing toys, and — the hardest thing in the world — sharing your parents' attention. I bought the book *Julius, the Baby of the World,* by Kevin Henkes, to give Mary and her children. My girls and I drove to the hospital to see Mary, Kaitlyn, and their new baby, Brent.

When I arrived, Mary looked pretty good and was sitting up holding the baby. I presented Mary with the book for Brent and Kaitlyn, then proceeded to sit on the floor, gather the girls into my arms, and read the book to Mary, baby Brent, and the whole family. Relatives arrived and came into the room. I think they thought, What is this crazy woman doing, reading that book on the floor? They all stood and listened as other family friends stuck their heads in the door.

If you haven't read this book, it is the story of a little girl mouse who has just gotten a new baby brother named Julius. She's not too happy about the situation, since he's too little to play with, he has stinky diapers, and everybody thinks he is absolutely precious with his cute little baby nose. And she stays unhappy, until they bring the baby mouse home and a cousin declares that he is too little to play with and he has stinky diapers. Well, the older sister mouse can't possibly have anyone say such terrible things about her new baby brother, who is absolutely precious with the cutest little nose.

The girls hung on my every word, and baby Brent got to hear his first book read aloud just hours after he was born. And I never really thought too much about that day, since reading is just what I do, until one summer day when Kaitlyn entered a Little Miss Jefferson beauty pageant.

Kaitlyn really seemed to enjoy Jefferson's beauty pageants, and she was the one who actually changed my mind about pageants after the traumatic experience I'd had in high school. Her eyes would quite literally light up and sparkle as she got into her pageant mode. My Madeleine used to do the little pageants, too, until one year she received first runner-up but didn't win first place. She stomped off the stage declaring, at seven years old, "What a rip-off! I looked like I was having a good time, I did the pretty feet, and I didn't mess up once. This pageant is rigged and I'm out of here!"

I couldn't help laughing, since she reminded me of what Scarlett O'Hara must have been like as a child: a pure dee drama queen. I sat her down later and told her that if that was going to be her attitude, then it probably was best that she didn't continue the yearly tradition. The pageant was supposed to be fun and help a little girl gain confidence in herself. Beauty pageants were not about being a winner above everybody else. I told her she needed to learn a valuable lesson in good sportsmanship and that she was old enough to know how to act like a good young lady.

Even though Madeleine never did another pageant, Kaitlyn continued, and, since she was Madeleine's best friend, we wanted to support her in this venture. We always came to see Kaitlyn. One year, when it was each girl's turn to walk the runway for the judges, the master of ceremonies read a list of their likes, favorites, and dreams. As Kaitlyn, a bright little beam of sunshine, started down the runway, he read, "Kaitlyn Whatley is seven years old and her favorite movie is *Josie and the Pussycats.* Her favorite food is macaroni and cheese, and the woman she most admires is Kathy Patrick."

I immediately looked at Mary and started crying. "Mary," I sputtered, "did you know she was going to say that?"

"Yes, Kathy, and it's true. She really looks up to you."

I realized right then and there exactly what a responsibility I have to live up to for this little girl. She is looking to me for what she thinks is important. I had no idea I was such an influence on Kaitlyn's life and, now that I know, I don't take this responsibility lightly. In that moment God was telling me that I have a purpose. I firmly believe that, when we are born, God has a plan for us.

To me, this was God whispering in my ear. I still get chills when I think of that day.

I also was more than a little bit thrilled and relieved when, the next year, Kaitlyn walked the runway and they announced, "This is Kaitlyn Whatley and the woman she most admires is her mother."

Me too, Kaitlyn, honey. Me too.

Mary continues to amaze me with her friendship. One day, I had stayed too late at my Pulpwood Queens of Squire Creek Book Club, clear over in Choudrant, Louisiana. I was going to be lucky if I got back to Jefferson in time to get everything ready for my Pulpwood Queens meeting. When I turned onto Highway 59, just fifteen minutes from Jefferson, I noticed that my damn minivan was on empty. I figured I had about thirty miles left in my tank before I actually ran out, so I didn't worry about it too much. But right at the city limits sign, my van started to sputter and start. The engine died. I could not believe it. I had run out of gas less than a mile from the gas station.

I coasted over to the shoulder of the road. As I rolled down the windows, the full extent of the summer's July heat came blasting in. My back instantly became wet from the humidity and heat. I reached for my cell phone, trying to think whom to call. Jay was out of town on business. Beverly, my best friend, aunt, and neighbor, had just left to take her daughter to basketball camp in Louisiana. Who wasn't working?

I couldn't think of a soul. Then it dawned on me to call Mary. She was still at home on maternity leave with Brent. Could she possibly run me to town for some gas? I gave her a call. Before I could even go into the details, she interrupted and said, "I'll be right there." And hung up.

Holding the phone in disbelief, I settled in for the wait as semis sped by, leaving my damn minivan — yes, that is what I called it — shaking in their wake. I was searching the glove box to see if I could find a napkin or a Kleenex to wipe the sweat off my forehead when I looked up to see Mary in her SUV making a U-turn right in the middle of the highway and pulling in right behind me. I got out, and as I approached her car she was already pulling a five-gallon gas container out of the back. By the time I got to her, she said, "Stay with Brent," and proceeded to walk to my van to fill it up.

I went back and looked in the car, which she had left running, and spied baby Brent safely bundled in the car seat asleep. I hollered, "Good grief, Mary. How in the world did you get here so fast with the baby and all, and where in the world did you get the gas can?"

"I just grabbed Brent, bundled him in his car seat, and stopped at my father-in-law's, and he had the gas."

She poured the last drop of gasoline into my van's gas tank. I could hardly believe how fast this girl worked. Obviously, she knew how to take care of herself and — thank God — me too!

"Mary, do you think the van will run? Do I have enough gas to make it to the station?"

"Yes, you'll make it."

As she walked back to her vehicle to put the gas container in the back, another semi went by the road a little too close for comfort.

"Mary," I hollered back to her, "follow me up there just in case. How much gas do you have?"

"I've got about a quarter of a tank," she yelled as she climbed into her vehicle.

"Okay, just follow me up there and I'll fill up your tank." She started to argue and I just gave her a look. "Come on, before I get run over by a semi!"

I hopped into my vehicle and the damn minivan started up on the first try. I turned on my signal to pull out and waited so Mary and I would both have time to pull out together. I was at the gas station in less than two minutes. As I filled up both our vehicles, Mary just stood to the side with her arms crossed and shook her head back and forth.

"Kathy, you should never let your car get that low."

"I know, Mary, but then I would never have found out what a great friend I really have in you."

And we both started laughing. I paid for the gas and Mary went on her merry way as I went on mine. Every time I think about how fast she reacted and was there for me — with a brand-new baby, no less — I just shake my head. Poor little Brent couldn't have been more than two months old!

One time Madeleine had asked Kaitlyn to go to the movies, but Madeleine had forgotten she had a Girl Scout meeting, so she couldn't go. Kaitlyn and Madeleine, though best friends, are one year apart. Kaitlyn was in the next-younger troop. Mary had already dropped Kaitlyn off at my shop. I apologized for both Madeleine and me, since we had both just let that meeting slip our minds. I asked Kaitlyn if she wanted to go to the show anyway with just me. She did and so we went. We had a great time, and I kept thinking: This is what it must be like to be a grandmother. I can tell you that I really,

really liked the experience. I just hope that I am alive to experience it with my own!

We talked about school and what she was reading, and she told me that she was still the top reader in her class and got straight As. The whole experience reminded me that, one New Year's, Jay and I had been invited to a New Year's Eve party and Mary volunteered to keep Madeleine. I thanked her but told her we probably weren't going to go, since Lainie wasn't old enough to stay by herself but was too old for a babysitter.

"Well, Kathy," Mary said, "Lainie can stay with us too. We aren't going anywhere and it will be fun."

So both the girls stayed with Mary, and Lainie had such a blast that it has become a yearly tradition to stay at Mary's on New Year's Eve. This past New Year's, we were on our way home from the holidays in Colorado and the girls did not get to stay with Mary after all. I don't know who was more disappointed, Mary or the girls.

I would like to say we have an equal fifty/fifty friendship, but I honestly think Mary is the more giving one. This past Christmas, when she found out we were going to be gone for the holidays, she and the kids came over for Christmas supper the night before we left. Mary surprised us by making all of us special polar fleece blankets to take on our trip. Mine was hot pink and leopard print, Lainie's was black on one side and had pink cool cats on the other, and Madeleine's was camouflage green to match her newly decorated camouflage room. There is nothing more special than a gift made by the hands of someone you love.

We treasure our blankets. They keep us warm at night but, more than that, they blanket us with Mary's love. Now I'm just trying to figure out how in the world I can ever beat that, and the fact of the matter is, I can't. I just know that I'll cherish my days and times with Mary, and look forward to watching her and her children grow into beautiful and outstanding citizens.

Mary and I don't talk to each other every day, or see each other every week. But when we do get back together, we always pick up right where we left off. If you have a friend like that, you ought to treasure her. I count Mary as one of the crown jewels in my crown of life.

Now I am no saint — far from it. Once, when the damn minivan was filled with what seemed like a hundred loud and rambunctious Girl Scouts, I started yelling and hollering just like I do at my own kids at home. The van got quiet and, after some whispering, my daughter Lainie said, "Mom, we have all taken a vote, and we feel that you have won the award for the most yellingest mom of all the mothers."

> Treasure your friends; they are like diamonds in your tiara. Treasure your family; they are your crown jewels.

Without skipping a beat, I hollered back, "You're absolutely right, children, and that is because who else's mother has you kids all the time? You all practically live at my house, so I see you as my own. What other mother has not one or two, but over forty slumber parties under her belt?" That shut them up for a minute. Then they all started snickering hysterically in the backseat.

I know that I am not perfect, but by God, one little girl looks up to me, and that is good enough for me.

I honestly can't think of anything more fun than sharing a good read with my children, their friends, and my friends. Here are some great books I have liked to share with my friends and my daughters as they have gotten older:

Another Song about the King **by Kath-**

ryn Stern Young Silvie lives in the shadow of her overly dramatic mother, whose one defining moment was that she had a date with Elvis as a teenager. A well-written look at a mother-daughter relationship that is a time capsule of the 1960s.

The Devil Wears Prada **by Lauren Weisberger** Smart yet unfashionable Andrea Sachs goes to work as the assistant for *the* iconic fashion editor Miranda Priestly. (The author is a former assistant to *Vogue* magazine editor Anna Wintour.) I must insist you see the movie too, as Meryl Streep as Miranda Priestly is sublime.

Fat Girls and Lawn Chairs **by Cheryl Peck** This is a wonderful and irreverent short-story collection the author originally wrote for her family and friends. My daughter Lainie and I spent a wonderful afternoon reading aloud the stories about Cheryl and her cat Babycakes.

Charms for the Easy Life **by Kaye Gibbons** In this book three generations of women live under the same roof. The story is told through the eyes of the granddaughter.

The Notebook **by Nicholas Sparks** Never in a million years would I usually pick up a book by an author this famous because I like to find the yet-undiscovered author.

But my friend and author Andy Behrman sent it to me for my birthday, and I have been a Nicholas Sparks fan ever since.

***Drunk With Love: A Book of Stories* by Ellen Gilchrist** Thanks to my fellow bookseller and friend Lori, I now have in my possession every Ellen Gilchrist book — and most of them are signed. Knowing how much I loved Ellen Gilchrist's writing, Lori gave me all her books as a going-away gift when I left the bookstore where we both worked as buyers. I will treasure those books and her friendship forever.

***Summer Gloves* by Sarah Gilbert** Pammy is former Miss South Carolina. Her mother was Miss New Jersey, and now Pammy has her own twelve-year-old daughter entering every pageant in sight. The grand finale has all three entered in a grandmother/mother/daughter beauty pageant. You've just got to love these beauty queens.

***What Comes After Crazy* by Sandi Kahn Shelton** Maz, who was raised by a sex-crazed fortune-teller, escapes into what she thinks will finally be a normal life: marriage. Her life soon collapses when her philandering husband has an affair with their daughter's teacher at the babysitting co-op where parents take turns babysitting each other's children.

CHAPTER 13
I LIVE FOR
ADVENTURE

"Life is either a daring adventure or nothing. To keep our faces toward change and behave like free spirits in the presence of faith is strength undefeatable."
— *Helen Keller, "Let Us Have Faith," 1940*

Every day of the week, my fellow bookseller Fred McKenzie comes riding up the sidewalk on his bicycle on his way to work. If I see him coming, I will go outside to greet him; I cannot wait to see Fred's smiling face. Fred is my hero.

Always in a good mood, Fred just makes my day. Of average height, slight in build, balding with a fringe of lightning-white silver around his head, he always wears a hat or cap that makes him look quite dapper with his silver beard and mustache. Seeing him run down the street — he is always forgetting a scheduled meeting — you would shake your head in disbelief to learn that Fred is eighty-eight years old.

> "The old believe everything; the middle age suspect everything; the young know everything."
> — *Oscar Wilde, Irish poet, novelist, dramatist, and critic*

Fred owns and operates Fred McKenzie's Books on the Bayou, which is directly across the hall from Beauty and the Book. Both our businesses are housed in a historic home, which was built right after the turn of the century. We share the bottom floor with a hallway dividing our shops. He carries more used books than you can possibly imagine. They run floor to ceiling on his shelves, and who knows what treasures lie within? Fred's old wooden desk faces his front door, and he is the first thing you see when you walk into his shop. That is, unless he is in the back, building additional shelving to house more books, which come in every day.

Once Fred stuck his head in the door and asked if I could help him out. Seems he'd bought out yet another used bookstore, and we were both just in our first month of business at our new locations. With books from floor to ceiling, Fred asked me, "Kathy, I have just bought out a used bookstore. I have twenty-two six-foot bookshelves being delivered full of books. Can you help me find a place to put them?"

Somehow Fred and I found spots for all those shelves, but to say that the store is cram-packed would be an understatement.

One day, when we had first opened, he came and took me outside to see his latest building venture. I was looking around for his project when Fred said, "No, Kathy — up there." As he pointed up, my eyes were drawn skyward to see a big hand-painted sign that read BOOKS in bold letters up on the roof.

"Good grief, Fred. How did you get that up there?"

"Well, I climbed up the ladder," he said. "I just hauled that sign up there and nailed it down."

That was a couple of years ago; he was only eighty-five then. Every day I am surprised at anything Fred decides he might do.

I would love to have Fred's qualities and zest for life. He is kind, always happy, and living for every hour, minute, and second. He is the author of many books, mostly on the history and genealogy of our area in Texas. He wrote *Avinger, Texas, USA* and *Hickory Hill* and is the town's historian. He has probably done more research and writing than anyone in this town on its genealogy and history. But the one Jefferson visitor whom Fred has researched the most would have to be the infamous Diamond Bessie.

> I have learned how to be a big talker by spending the formative years of my life being a big listener. Fred and I also have that in common: he's a big talker today, too!

Diamond Bessie's real name was Annie Stone, and she was known for her good looks, fashionable dress, and lavish diamond jewelry. Bessie was one of her pseudonyms; she had quite a few. She ran off from her home in New York state as a young woman and soon acquired quite a reputation. She had many benefactors, and they were known to give her gifts of diamonds. She wore those diamonds everywhere and became known as Diamond Bessie. Who was this mysterious woman? And what had she done to be given such extravagant gifts? Fred has spent many years trying to figure exactly who she was. We always joke back and forth that Diamond Bessie had to have been the original Pulpwood Queen for her love of diamonds. And I love to hear Fred recount her infamous legend.

> "Big girls need big diamonds."
> — *Elizabeth Taylor*

The Jessie Allen Wise Garden Club hosts our local Annual Pilgrimage and Home Tours. Women and men dress in period costumes and open their private historic homes for tours during an afternoon back in time. The garden club also coordinates the Excelsior House Players, who put on the now-famous *Diamond Bessie Murder Trial,* a reenactment of the actual case that set new precedents in Texas legislature.

The setting is historical enough, but who has played the role of the gravedigger in the play for over thirty years? Fred McKenzie, that's who, for who else could be more perfect for the part? He knows the history of our town backward and forward.

Not a day goes by that somebody doesn't stop by to see Fred and ask about his or her family's history and genealogy. One day it was Don Henley of the Eagles fame. Fred is also the keeper of many a family secret, like who had children by whom in this black-and-white community. All the other secrets lie buried now in the Oak Forest Cemetery, including the body of the murdered Diamond Bessie. She lies beneath a monument that has no history. Evidently, in the late 1920s, members of the cemetery committee noticed that a new monument had been placed there, one that read simply, BESSIE MOORE, 12-12-1876. The earlier monument marking her grave was gone.

A couple of summers ago, Fred traveled to New York state and found the house where Diamond Bessie grew up. Being the adventuresome lad he has always been, Fred met the surrounding business owners and became acquainted with them. He has the photos to prove it. When he came back, he couldn't wait to show me the photographs he'd taken of Diamond Bessie's home.

"Kathy, the neighborhood hasn't changed much in the hundred-plus years since Diamond Bessie grew up there. It's a sorry neighborhood on the downside."

"Is this a bar, Fred?"

"Oh, yes, that little ole gal you see in the photo was the bartender, and she was more than helpful in talking about the neighborhood."

As I looked at the woman, with her missing teeth and her arm slung around Fred in the photo, I stifled a giggle and almost choked on the next photo. "Jiminy Crickets, Fred, is this what I think it is?" I asked as he quickly grabbed it away.

"Well, Kathy," he explained nervously, "I guess you could say there was a pornography shop right next to the bar. The lady who worked there was very cordial and — yes, I guess I better not show this picture to too many people."

"I would probably set that one aside, too," I told Fred about the next shot, "because I'm not sure the citizens of Jefferson are ready to see you and shelves lined with battery-operated dildos in the same picture."

> "Diamonds are forever. E-mail comes close."
> — *June Kronholz*

We both chuckled and laughed, since Fred, even though he is the most Christian man I know, is not a prude. He was just really trying to find out about Diamond

380

Bessie and the neighborhood where she used to live. He would have made an amazing anthropologist: just the facts, ma'am, regardless of the dire circumstances and state of affairs.

Fred had taken photos of the house where Diamond Bessie had lived from all angles. I cannot even imagine what people must have thought of this eighty-plus-year-old man tromping through the weeds and such, snapping photographs with his camera — it just makes me smile. Yes, he is my inspiration.

Fred and I talk about books a lot, and one day, as I was spending some time going through his stacks, I found a book titled *I Married Adventure: The Lives of Martin and Osa Johnson,* by Osa Johnson. I ran to Fred's desk to show him and tell him that I had found the same book at a neighboring antique store a couple weeks before. The cover here was a faded zebra print just like mine and I encouraged Fred to read it, because, just like the author and her husband, Fred was a pilot. In fact, he is a member of the United Flying Octogenarians.

Fred entered the Naval Aviation Cadet Program in August 1941 and soloed the following September, then transferred to Jacksonville, Florida, for advance training.

He is still flying today. Now, jokingly, I tell everyone that the reason he rides his bike back and forth to work every day is that he does not drive too well in reverse. In fact, he hits things — mostly other cars. The reasons he is still flying are: first, he passes his physical every year, and, second, you cannot go in reverse in a plane — though I bet Fred could figure out a way.

Fred has always been adventurous, so when he finished reading the book we had lots to talk about. Osa and Martin Johnson hailed from Kansas and were among the first adventurers to visit exotic locales and photograph and film the native people. Their stories had me on the edge of my seat, because, like one of my favorites, Tarzan, these people lived for adventure. Both had become pilots and were the first ones to photograph Mount Kilimanjaro. To be the first to fly and see such an incredible vista — I just could not imagine how incredible that feat alone must have been. Fred was hooked. He had to read the book. I left him at his desk reading.

The next day, Fred excitedly called me into his bookstore. He had done a little research and found that the hardback edition of this book had a first printing of only 400. We both had hardback books. A book in fine condition was valued at $500. We were both delighted at such a find. Would I ever sell mine? No way; I love this book. Would Fred sell his copy? All I can say is, make him an offer. There is nothing in the world Fred loves more than to deal in used books.

In his shop, Fred has wonderful photos of his life's adventures. One photo I am particularly fond of features Fred as a young man with his shirt off, wearing khaki shorts, a pith helmet, and hiking boots, standing with a bunch of natives in the Solomon Islands off the coast of New Guinea. They are all holding spears, and Fred is smiling for the camera. I am sure his wife, Ann, would also have said, "I married adventure," when she married Fred.

> "A man practices the art of adventure when he breaks the chain of routine and renews his life through reading new books, traveling to new places, making new friends, taking up new hobbies, and adopting new viewpoints."
>
> — *Wilfred Petersen*

One day, Mary Hileman, a member of the

Jessie Allen Wise Garden Club, came over to chat with me. Fred had just left on his bike and was going close to fifteen, maybe twenty, miles an hour down the sidewalk when Mary spotted him.

"My lord, look at Fred pedal that bike," she said. "He is the fastest man I have ever seen ride a bicycle."

"Well, Mary," I quickly replied, "you do know he has a motor on that thing. When he gets to pedaling, the motor kicks on and away he goes."

And I swear to God Mary said, "No shit! I thought the little guy was pedaling that damn thing and in the best shape."

Two tourist bystanders who overheard the whole scene just cracked up laughing. I went on to tell them that one day Fred was headed home on his bike and took off up a hill. As he rode, he passed a fifteen-year-old girl who was huffing and puffing. He told me he looked back and saw that she was totally disgusted with herself. Fred got quite a chuckle over his one-upmanship on the bike. We were all laughing by then, so I told them a few other Fred stories.

One was the time he flew his plane into the airstrip he built out at his place on Lake O' the Pines. A certain black dog always came out barking at the wheels of the plane when he would try to land. Fred said that he was so busy trying not to hit the dog with his plane that the wing tipped the edge of the trees, and dang if he didn't total his plane as he crash-landed.

Another story is the time I noticed a new photo in his shop, which looked as if it were taken a few feet away from the top of our building. I asked him, "Fred, was this taken from a plane? This photo looks to be an aerial view of our shops."

"Oh, yes it was," he said. "I took it."

"But, Fred, who was flying your plane when you took it?"

"Oh, that was me, too. I just leaned out when I flew over and took the photo."

"When was that, Fred — several years ago?"

"Oh, no, I took that just the other day."

Fred was eighty-six at the time.

Fred has spent his whole life documenting other people's lives and family histories. Though those books are fine and entertaining reads, I would really like to read a book about Fred and his life. My husband, Jay, was talking to Fred one day when Fred told him that, from 1943 to 1948, he documented life in and around Marion County. He traveled up and down its roads filming as he drove. He went into every business and documented just an ordinary, typical day of business in Jefferson. Jay took those films and transferred them to disc so they would not be lost. We were visiting Jay's grandmother one Easter and watched the CD after our Easter dinner and egg hunt. I watched from the car window as Fred drove down the old Karnack Highway that ran right through Jefferson by the old Texaco station. Through Fred's film, I walked through the doors of the drugstore and the car dealership, and watched uniformed attendants running to wait on customers. Wait a minute — employees dressed in starched uniforms with caps, running to wait on customers? I was dumbfounded. I watched as we went from business to business, and the stores were crowded. I have never seen that many people in a furniture store on a weekday, not even during a going-out-of-business sale.

I am sure my mouth was hanging open. Seeing children playing boisterously on the old Jefferson High School playground was surreal. Most of these people were no longer even living. I was taken back in time. The Smithsonian needs to see this film — it is the history of East Texas live on celluloid.

The scene that made me understand how different East Texas was back then was the film Fred took of an all-black traveling minstrel band marching through downtown Jefferson. Black women were on the street holding the hands of their little white charges. One black woman who had a white baby on her hip was so overcome with the joy of the music she began to dance in the streets. I had never seen so many black people in town on a normal business day.

Fred and I have had many talks about the black history versus the white history of the area. Not a week goes by when someone (black or white) doesn't come in to talk about their family history. Fred is Jefferson's own living treasure. He knows all the facts. We can all learn from a man who has been a lifelong learner and reader.

"We're all made up of stories. When they finally put us underground, the stories are what will go on. Not forever, perhaps, but for a time. It's a kind of immortality, I suppose, bound by limits, it's true, but then so's everything."
— *Charles de Lint, Celtic folk musician and storyteller*

Recently I found out that Fred wants to go back to New York because he has discovered that Abe Rothschild, who was found guilty of murdering Diamond Bessie, is buried in Brooklyn. However, his family feels that he has reached an age where he should not travel by himself. Therefore, we have made a pact. If at anytime I have to go to New York regarding this book, Fred is going too. We are going to find Abe Rothschild's grave. I can hardly wait for that day, because I can think of nothing more fun than going on an adventure with Fred McKenzie.

When we celebrated Fred's eighty-eighth birthday I asked him if he would like to go as my date to hear Michael Martin Murphey, who was cutting a live CD at Music City Texas. Fred was thrilled to go! I picked him up at 6:00 p.m. at the shop; he was sitting out on the front porch ready to go. As we drove to Linden, he drifted off to sleep and I let him. It was to be a big night ahead and I knew that he had not had a nap that day because it was the weekend of the annual Boo Benefit Motorcycle Rally, an event where thousands of motorcycle enthusiasts came to Jefferson to partake in all kinds of activities with all money raised going to the burn center in Shreveport. Our town was full of motorcycles parked as far as the eye could see in all directions. I was sure he was worn out from all the festivities.

The minute we pulled into Music City, Fred was wide awake. As we entered the building everyone greeted us, wishing Fred a happy birthday. I had told a few folks ahead of time that this was Fred's special day. Our friends Jean and Russell Wright met us at the ticket booth where I was to pick up and pay for our tickets but we were ushered right in because the Wrights had already paid. I had intended to treat Fred, and Jean and Russ ended up treating both of us. I was dumbstruck. I started to argue, but they were smiling ear to ear and told us to go find our seats.

Fred and I entered, and I asked Fred if he wanted some of the best chicken and dumplings or maybe a barbeque sandwich before the show. He opted for the barbeque, and we sat at a little table behind the sound booth to enjoy our meal. Matt Early, who ran the sound board, immediately came over to hug us and wish Fred a happy birthday. Pat Robinson, who plays in the band the Moon & the Stars, brought us another surprise: VIP passes to go backstage after the show and have some more fine food and hang with Michael Martin Murphey. I'd wanted to surprise and treat Fred, and I was surprised and treated too!

Richard Bowden, the president of Music City Texas, came over with birthday wishes and then whispered to me, "We will mention Fred's birthday during the show." I nodded, smiling, as Fred took another bite of his barbeque. Lots of folks drifted over from Fred's hometown, Avinger, Texas, to say hello.

The lights flickered and it was time to get our seats. Ours just happened to be front and center. When Michael Martin Murphey came out, I felt just like we were having our own private concert. The music was fine and the show very personal. I looked over at Fred and I could tell he was thoroughly enjoying himself. During the intermission, Richard Bowden came out and recognized a young man who was on leave from Iraq and soon going back. Then he told everyone there was an author and historian here who was having his eighty-eighth birthday, and could everyone please give a big hand to Fred McKenzie? A spotlight shone on Fred as he stood up and took a bow. I thought, What a nice thing for Music City Texas to do. I know I will never forget that moment.

As Fred and I drove home that evening there was no dozing off this time. We chatted all the way about the turn of events, and how we were treated like Kings and

> You want to know what love is? Get off
> the interstate and go visit a small town.
> Smile and wave to each passerby and
> next thing you know, you will be em-
> braced. Eat at a mom-and-pop restau-
> rant; ask, "What's your best dish?"
> Walk the streets and visit the shops.
> There is nothing better to me than
> small-town life.

Queens. I know Fred had a good birthday. What I did not expect was just how much fun I would have hanging out with Fred at Music City Texas. As I dropped him off (at 1:00 a.m.) at his home where he lives with his daughter and son-in-law, Carol and Paul Harrell, we laughed at the late hour. I wondered if Carol had ever dreamed that, as a grandmother herself, she would be waiting up for her father on a date! Fred and I will be laughing about our evening out for years to come.

One December, Fred convinced me that I needed to ride with him on his two-seater bicycle in the Jefferson Christmas Parade. I told him I would, not really thinking about what that would entail. I helped Fred by weaving red tinsel through the spokes and around the handlebars. We put several copies of the books he had written, *Avinger, Texas, USA* and *Hickory Hill,* in the basket with a sign on the front reading, FRED MCKENZIE, AUTHOR. I duct-taped two flashlights to the handlebars so we would have some light, since the parade is always held after dark. On the back, I duct-taped a plush toy of snowmen that when pressed played music. Fred wore my red Santa hat trimmed with white fur, and I wore my red furry cowboy hat with white-fur trim — and both of us wore our Christmas sweaters. We were ready to go.

Fred was going to ride on the front and I was at the back. I thought I had the easier position — all I had to do was pedal and ride. But as we started, I realized that riding this bike was going to be like steering cats on a cattle drive: the whole dang contraption just kept floating back and forth across the road. Since I had no control over the steering, I spent most of the ride yelling, "Watch out, Fred! Don't hit that —" as we narrowly missed a float or a person scrambling to get out of our way. To say it was a wild ride would be putting it mildly; I was hanging on for dear life. Sweat was running down my back. I wasn't sure if it was from the sheer exhaustion of pumping the pedals of that steel death trap on wheels, or from sheer panic that we nearly rear-ended the numerous floats and children scrambling for the thrown candy and treats that, unfortunately, had gotten in our path. The only way to stop was to drag our feet. Shoot, Fred's feet could not touch the ground when we stopped, so the full weight of this titanic steel suicide machine rested completely on my thighs. My legs have a hard enough time keeping up with me, let alone Fred and the sixty tons of metal we were riding. I know you're thinking that I'm exaggerating, but in all the thirty or forty years Fred has ridden that bike in parades, I have noticed in photographs that he never had the same woman on the bike twice. Just think about that, will you?

When we slid back into the yard of our bookstores, Fred hollered back, "Ready to go around one more time?"

I am huffing and puffing to catch my breath and ole Fred is just looking at me with not a drop of sweat on his brow. In fact, I believe the man is in better shape than anybody I know. As I stumbled off the bike, exhausted, now completely drowning in my own sweat, I looked up at Fred and asked him, "Are you absolutely out of your mind?"

> "Literature is my Utopia. Here I am not disenfranchised. No barrier of the senses shuts me out from the sweet, gracious discourse of my book friends."
> — *Helen Keller, American memoirist, essayist, and lecturer*

He laughed all the way up the steps while we struggled to get the bike back on the porch. He was a-chuckling to some nearby tourists as I ran in the door to find water. I was parched beyond belief.

Fred not only runs a bookstore, but he also is quite the thespian. Ever since the Excelsior Players recently extended their theater repertoire to Dickens's *A Christmas Carol,* Fred has portrayed the character of Jacob Marley's ghost. First gravedigger, now ghost — I am afraid Fred is getting typecast. When they asked me and my girls to join that production, I agreed so I could be on the same stage with Fred. I was not disappointed. He makes a great Jacob Marley's ghost. There is nothing Fred could ever tell me, either from his past, present, or future, that would ever surprise me. I just know that there is a lot to learn from Fred and I am looking forward to our New York days.

Author Willie Morris, whom I have always

loved, wrote a book called *New York Days.* It is a story about a Yazoo City, Mississippi, boy trying to make his mark in the big city. It's a great book and everyone should read it. Nevertheless, I don't think old Willie has anything on Fred. Now I just need to convince him that, instead of writing books about others, he needs to write a book with me, called *I Lived for Adventure.* At least that way we could get Fred's own story down on paper: the story of author, historian, actor, pilot, and friend.

> Life is an adventure; make adventure your middle name.

The Tarzan books I read when I was a kid lifted me out of the wheat fields of Kansas and planted me in a lush African jungle. Here are more books that will take you places you have longed to visit:

West with the Night by Beryl Markham Though the author was born in England, she was raised in East Africa and ran with the natives. She then followed in her father's footsteps, training and breeding racehorses. I was fascinated by how she became an African bush pilot and then went on to be the first person to fly solo from east to west across the Atlantic.

Out of Africa by **Isak Dinesen** This is a beautiful tale of a woman who once owned a farm in Africa that was later made famous by the film. Both the book and the film had me totally mesmerized from start to end.

The Canal House by **Mark Lee** This is the story of four people, three men and one woman, whose lives become entwined in the international setting of civil unrest from Uganda to East Timor. If you have ever wanted to know what it is like to cover the world's news as a foreign correspondent or photographer, you will learn firsthand in this edge-of-your-seat love story.

Travels with Myself and Another by **Martha Gellhorn** This is Martha Gellhorn's memoir of her life as a foreign correspondent. She also just happened to be the third wife of Ernest Hemingway. She was fearless, and her story is a must-read for any adventurer.

Coming of Age in Samoa by **Margaret Mead** When I took my first anthropology course I fell in love with the stories and books of Margaret Mead. She was a woman way ahead of her time who still fascinates me to this day.

I Married Adventure: The Lives of

Martin and Osa Johnson by Osa Johnson This small-town Kansas girl went on to be the first to film the wild tribes of Borneo, gorillas, cannibals, and Mount Kilimanjaro. Osa lived a life the likes of which will probably never be lived again. I get that same pump of adrenaline just talking about her adventures that I had when I first read the book.

Last Moon Dancing: A Memoir of Love and Real Life in Africa **by Monique Maria Schmidt** The author grew up in a Mennonite community in South Dakota, then after graduating from college headed to Africa with the Peace Corps to teach the native children. A wonderful read for the adventurer and also for anyone who truly wants to teach abroad.

Honeymoon with My Brother **by Franz Wisner** A man gets dumped just days before the big wedding. He has already paid for the honeymoon, so he invites his brother, recently divorced. Fifty-three countries and two years later, we hear their tale that begins with sadness and ends with brothers who become reunited with many adventures yet ahead.

The Poisonwood Bible **by Barbara Kingsolver** When I found out as a child that I could not become a nun since I wasn't Catholic, my mother suggested I become a missionary. Here is the story of one Baptist preacher who drags his four daughters and wife to the Belgian Congo in Africa, where everything pretty much goes to hell in a handbasket.

Without Reservations: The Travels of an Independent Woman by **Alice Steinbach** Pulitzer Prize–winning writer Steinbach takes a sabbatical from the *Baltimore Sun* to travel around Europe. This is one thing I have always dreamed of being able to do, and through Steinbach's book I can travel along.

CHAPTER 14
GONE WITH THE
WIND — LITERALLY!

"Nothing can be so perfect while we
possess it as it will seem when
remembered."
— *Oliver Wendell Holmes, Sr.*

My forty-ninth birthday hit on a Saturday
with a big storm brewing outside my shop.
As I was working, some of the Pulpwood
Queens stopped by to drop off their gifts.
From Auntie Bev I received a leopard hand-
painted martini glass, complete with the
recipe for a Leopardtini painted on the bot-
tom. From my good friend and financial
adviser Pam McGregor I received a wooden
aqua-painted desk sign that read YA-YA.
Sandra Phillips and Jean Maranto gave me
a hot pink sign that said DIVA in resplendent
jewels hanging from a chiffon ribbon. Cards
and flowers poured in — everyone knows
how much I love to celebrate birthdays!

I shrieked and eeked as I opened each of

the packages, tossing ribbons and paper aside fast and furiously. It looked like a tornado had hit the shop, and in reality a storm was coming and coming fast. Dark clouds had rolled over our neck of the woods. Big splats of raindrops hit the shop, sounding like war drums as they battered the roof and windows. As bad as it looked outside, this was nothing compared to what was happening down on our Gulf Coast. Hurricane Katrina was on its way, and this storm was one for which we were totally unprepared.

> "Yesterday is history, tomorrow is a mystery, today is God's gift — that's why we call it the present."
> — Joan Rivers, American humorist

As America watched, fascinated by this tropical blast about to hit our Gulf shores, we wondered what was going to be damaged. New Orleans, a favorite place for the Pulpwood Queens and Timber Guys to visit for some downtime, was a major concern. I had spent many a fall weekend in New Orleans attending the Mid-South Booksellers Convention and had author friends and favorite stops that kept that coastal town near and dear. We all were expecting a big storm, but what happened was something no one in the South, or really in America, was prepared for.

As we worked in the salon that week, all our eyes were glued to the television for the latest news on what was happening mere hours away. You could not pull me away from the television screen except to check the latest news on-line. When it seemed that the devastation of the storm had passed, we received word that levees had burst, flooding New Orleans. One of my best friends, Pulpwood Queen Carol Lancaster Lucky, had just signed a contract for a street-level apartment in the Quarter. Before she'd even had time to take up residence she was in fear that her new townhouse had flooded and the waters were quickly rising. People who were without transportation were trapped in this bowl twenty feet below sea level. Prayers were sent while we watched in horror as those stationed at the Superdome were left with parts of the roof of the dome exposed and no food or water or restroom facilities available.

I knew I was watching the biggest human tragedy that had ever affected my place in the world. Having lived in California, I still shiver every time I see the film clip of the car falling fifteen feet on the bridge when California had its giant earthquake. September 11 will always be ingrained in my memory because I had spent so many days on top of the World Trade Center for book publishers' sales conferences. I will never forget that I was cutting my friend Carol Harrell's hair when I received the call that the World Trade Center had been hit. I remember this because, for me, the feeling is exactly the same as remembering what I was doing when John F. Kennedy was assassinated, then Bobby, and then during the tragic assassination of Martin Luther King, Jr.

I live in a town where history is revered. Now I was being thrown full blast into one of the most tragic natural disasters in American history, and New Orleans was in my own Southern backyard.

People began to flood into our town. There were displaced families, first seeking shelter from the storm, then stuck because they had no home to go home to. We got the lucky ones, the middle-to-upper-middle-income families who arrived in their SUVs with vacation clothes, T-shirts, and flip-flops, thinking they would come to Jefferson for a weekend getaway and ride out the storm. They brought their pets and stayed in our many bed-and-breakfasts and hotels, many whose owners gave them special rates because they knew about their unique and tragic situation. They came for what they thought was two days, maybe three, a little mini-vacation.

Maybe they had grabbed the family picture album or a cherished memento, but most saw this as a short-term break from the coast. At first they sought out the sights and attractions of the town, riding the train, taking boat rides, or walking with the family on our ghost walk. Then reality set in. Their homes were completely flooded or, in some places, wiped from the face of the earth. No time frame was given for when the levees could be repaired. Water was still rushing in. We watched in horror as the people at the Superdome pleaded for necessities. There were rumors of rape and murder at this place of refuge, and we were horrified for those people.

Suddenly, folks began to see a clearer picture. They could not go back to New Orleans or the other coastal cities. Their money was running out and credit cards were being maxed. Their resources were spent, and all they had in vast supply was time — time spent waiting until they could go home. For some, their homes were gone, their jobs were gone, and these people were in shock.

They parked on my porch because I had placed my regularly delivered papers, the *Dallas Morning News* and the *Marshall News Messenger,* out there for them to read. Since I have wireless Internet service at my shop, some sat on the porch with their laptops, trying to reach family members. Playing cards were left on the table for them to pass the time. I placed a cooler of ice and jugs of lemonade out on the porch. I asked all the Pulpwood Queens to bring by fresh-baked treats. Then we put out a big sign that said FREE, HELP YOURSELF! I saw that they had all the time in the world and little else. What to do!

In times of distress, I always turn to books. I began to gather up a stack of children's books, books that make a great read-aloud, and sent out a press release to my Pulpwood Queens e-mail list. I didn't want these people to think I was offering charity, but the word got out, thanks to our area papers, and people began to come.

Years ago, my daughter Lainie had started a Splinter chapter of the Pulpwood Queens. Part of their literacy mission is to have story times for children. Lainie gathered up some of her favorite books and I watched my daughter read aloud to the children of the displaced families. I noticed that the parents, too, got caught up in the stories. She read *Weslandia,* fairy tales, and *Gilbert de la Frogpond,* and, as the children listened in rapture, I watched the adults relax and sink back into the cushions.

I pulled in Chisato Sayama, our exchange student from Japan, to teach the children origami. I would go outside and read aloud from *Old Dogs and Children* by Robert Inman, *A Change of Heart* by Philip Gulley, and *Life Is So Good* by George Dawson. No one seemed in a hurry, time passed, and everyone was able to escape for a moment the utter stark reality of being homeless.

I ran over to our nearby restaurant, the Hamburger Store, on one of those days to grab a bite and noticed my minister, Polly Standing of the First United Methodist Church of Jefferson, waiting tables. I asked her, "What is going on, Polly?"

The place was packed and I didn't recognize anybody as I watched them spoon up red beans and rice. "Kathy," she said, "the Hamburger Store has been feeding the displaced families for free, and I decided to chip in and help. Most of these people are stuck here with just the clothes on their backs. We're changing out the Sunday

405

school rooms tonight for emergency housing at the church."

As I watched her run from table to table pouring coffee and tea, I looked around and thought, What can I do?

When they gave me my to-go bag, I paid and ran back to the store. I gathered up every Pulpwood Queens T-shirt I had for sale in the shop. I sacked up the stacks of Kimberly Willis Holt's children's books that I keep in large quantities in my shop and trucked them all back over to the Hamburger Store. I walked up to Polly and told her, "Polly, here are some T-shirts and children's books. Would you please give them out to any of these families who need a clean shirt or a book to read to give them something to do with their children?"

"Sure, Kathy. I'll make sure that these go to those in need."

Then I asked her what time they were working up at the church to change out the rooms, and she told me about 6:00 p.m. I told her I would be there with bedding I could donate.

After work, the girls and I went through our closets and gathered up all the sheets, pillows, and bedding we were not using. Then I told the girls to go to their closets and drawers and bag up anything that no longer fit or that they couldn't or wouldn't wear. We lugged all the stacks of bedding and trash bags filled with clothes and shoes out to the car. It was packed as we drove to the church.

When we got there, Polly Standing and the church secretary, Paula Youngblood, were already at work. They had cleared out the rooms and were putting out air mattresses. Church members Jim and Kim Gallant had gone to Wal-Mart and bought every full-sized camping air mattress they had and donated them to the church. Others started arriving to help, and in no time we had each Sunday school room ready for a family for shelter from the storm.

> "To love what you do and feel like it matters. How could anything be more fun?"
>
> — *Katharine Graham,*
> *American newspaper publisher*

Pulpwood Queen Kay Brookshire approached me around this same time; she had a whole family staying with her. Other Pulpwood Queens had taken in members of a family: the mother and father had lost their home, their son and his wife had lost their home, and their daughter and her husband had lost their home. They had all gotten out of New Orleans with only the shirts on their backs and a baby car seat. Their twenty-two-year-old daughter was due with her first baby in mere weeks.

The Jessie Allen Wise Garden Club, Kay informed me, had decided to give this new mother-to-be a baby shower, and would the Pulpwood Queens be willing to help?

I put the word out to the Queens, and then my good friend author Andy Behrman e-mailed me from Los Angeles. What could he do? His wife, Julia Eisenman, had recently had their first baby and they wanted to help. I e-mailed Andy and told him that if they seriously wanted to help, they could provide some things for this new mother. While I asked the Pulpwood Queens if anyone would like to give a cash donation, Andy went to work in Hollywood. He and his wife collected money among their peers and coworkers and bought a Beverly Hills drugstore out of baby goods. They also talked the owners of their local baby boutique, Petit Trésor, into making a generous donation of a first-baby Moses basket filled with the most darling baby items. Two large crates of boxes were delivered to the shop for this new mommy's baby shower.

I am happy to announce that she had a boy and had more than everything she would ever need to welcome her new baby. She told me that, after her family lost everything, she was not very excited about having her baby. After the garden-club shower and all the generous blessings, she recaptured the thrill of having her first baby. The family has since relocated to Shreveport, found jobs, and are now beginning their new lives.

> "Only a life lived for others is a life worthwhile."
> — *Albert Einstein*

If I had not been a reader, I never would have met Andy, and his wife would never have been able to help this young mother and her family. Books take you to fabulous places. Even I had never dreamed that, as readers, we would ever come together in this way to help others in need. Andy and his wife have since then given generously to my church to help replenish dwindling emergency supplies, just in case this ever happens again.

I realized after this event that I treasured my books too much. I had built a house of books, and here were people who were living day to day, hand to mouth. I still love my books, but I will never look at tangible things in quite the same way. What really matters in life is that we love each other and that we have relationships that matter. As much as I love my books, there are other things that are much, much more important.

> Put God first, family second, then your friends, followed by good work. All the other stuff we accumulate in life is just stuff. Life is about loving one another and about our relationships.

Throughout this time, I gave away many of my personal items. You know, I haven't missed one single thing I gave away. A former minister, Jerry Vickers, gave what I think was one of the most eloquent and memorable sermons of all time called "Stuff" one Sunday in church. He said, "We buy stuff. We get so much stuff, we buy containers to put our stuff in. What is all this stuff, anyway? If we invested as much as we do in procuring stuff into our relationships with church, family, and friends, I assure you this world would be a better place."

Think twice about what you really need. I

read a book now and, unless I am going to read it again, I pass it on to my church, or school, or library, or family, or friends. Books are the gifts that keep on giving every time someone reads the pages.

Many people think that the hurricane disaster is over, but let me assure you it is not. People who believe it's over obviously have not been to the coast and seen the devastation. What happened on our coast is the nation's largest natural disaster ever. I think, as Americans, we sometimes forget the impact of such an event. It is out of sight and out of mind. But you all can do something to help; I suggest helping Louisiana libraries replace the books that were lost in the devastation or contributing to the Red Cross.

I turned fifty as I was finishing up this book, a year after Hurricane Katrina. I thought of all the birthdays I had had and what I wanted to do for this milestone occasion. Then it dawned on me: The first half of my life I had taken from the world, so the last half of my life I wanted to give back. I wanted to go to New Orleans. I wanted to see what happened down there. I got my wish and I went to New Orleans.

Jay and I traveled to the Big Easy with our good friends Carol and Randy Lucky. We were to be staying right in the French Quarter. We left after work on Friday, so it was late when we drove into New Orleans. I caught glimpses of FEMA trailers crammed into what looked like industrial parking lots. I could see traces of the high mark of the waterline that was left on the concrete barriers on the side of the freeway.

When you arrive in the Quarter there is nothing in the world that can describe the scent and heavy, wet weight of the air you breathe. New Orleans is a place where you live and breathe the history. The smell of the quarter is a perfume composed of old buildings, incense, sweat, urine, pure humanity, and delectable aromas of food wafting through the air. I love that smell. It not only gets in your clothes, but fills every pore of your body.

To think of the devastation of the great city known for its history, people, food, art, antiques, and culture is almost more than I can bear. New Orleans is a city that was made for a writer. Everywhere you look, a story has unfolded or is unfolding. To me, New Orleans is a woman who has fallen but still paints herself to regain her beauty. Each fold of her tawdry velvet dress reveals yet another hidden secret, a secret key to her past. She may appear worn and dirty, but if

412

you really look into her eyes, you see that she was once a great beauty.

During our trip, I awoke at the crack of dawn to see broken tiles on the neighboring roofs. That seemed to be the worst of the damage as far as my eye could see. The Quarter itself seemed in good shape. According to the natives, it was the outlying parishes that had suffered the worst of the damage.

We spent most of our time in the Quarter my birthday weekend, walking around and checking out the local color and eateries. The fare we tasted did not seem to have suffered from the storm — in fact, our dinner at Muriel's on Jackson Square was the finest meal I have ever been served. Sunday morning we went to the Katrina anniversary service at the St. Louis Cathedral, which was majestic and humbling. The massive organ had water damage from the hurricane, and now they would be raising the funds to help repair that damage.

At my birthday dinner I cried like a baby. A couple of wonderful old gentlemen sang "It's a Wonderful Life" to me at the Rib Room, and then "Happy Birthday to You" as I sat with my family and friends. I will never forget that moment and how thankful I was to have my loved ones with me on that special day in a city that had suffered so much.

What can we do to help this city? I believe the best thing you can do is to visit it to see for yourself. And there are more causes to choose from than you can imagine if you really want to help. I organized the first-ever Christian and inspiration book festival, Books Alive, which was a fund-raiser to help the church in my hometown that helped coordinate all the Hurricanes Katrina and Rita displaced-family relief efforts. We raised around $3,000 to help the First United Methodist Church in Jefferson for its mission and outreach programs. We still have some of the displaced families here; they have made Jefferson their new home.

There is much we can do, and it doesn't always have to be in the form of financial donations. We can give our time and our hands to work. There is nothing more important in this world than to serve others. Pay it forward.

When I first began reading books I looked in the pages and stories to learn. I have been a true reader for at least forty years of my life, and now when I read I look outside of my world. Reading can open your eyes to many things, and I happen to believe that reading also makes us better people and citizens.

The following are great reads that make you realize that every day, somewhere in the world, someone is dealing with a catastrophe, whether man-made or nature gone awry.

The Wonderful Wizard of Oz by L. Frank Baum You have to love a story about a Kansas girl who gets to go over the rainbow and meet some characters who change the way she looks at life altogether. Sometimes we have to go through a great storm to fully appreciate that "there's no place like home, there's no place like home."

Hatchet by Gary Paulsen A young man survives a plane crash in the Canadian wilderness with only one item to help him in his survival: a hatchet.

Eastern Sun, Winter Moon: An Autobiographical Odyssey by Gary Paulsen Paulsen recounts the horrors of his childhood when at the age of seven and during WWII, he traveled with his mother by ship to the war-torn and battle-scarred Philippines.

Hurricane Season: Stories from the Eye of the Storm by Karen Bjorneby The daughter of an air force pilot, Karen traveled the world and has now given us twelve stories that are compelling reads with surprise endings.

Where the Heart Is by **Billie Letts** Novalee Nation is headed down the highway, seventeen, seven months' pregnant, thirty-seven pounds overweight, when she loses her flip-flop (it falls out of a hole in the floorboard of her boyfriend's wreck of a car). Her boyfriend stops at Wal-Mart for her to buy some new shoes and when she comes out, he's gone.

Old Dogs and Children by **Robert Inman** Bright Birdsong's grandson is dumped at her door when her daughter takes off to get married again. Her son is caught in a scandal as he's running for governor of the state. Then, when it seems that things can't get worse, they do.

CHAPTER 15
PULPWOOD QUEENS
DO EUROPE

"I never travel without my diary. One
should always have something
sensational to read on the train."
— *Oscar Wilde*

After I discovered reading as a child, I also
discovered that books could take you to
unknown places and foreign countries. I
loved learning about India in *The Little Prin-
cess* and *The Jungle Book,* or about the
Swiss Alps in *Heidi,* or about England in the
Beatrix Potter books and *David Copperfield.*
But there was one place that was absolutely
magical to me, and that was Paris, France.

I became obsessed with France because of
the stories my French teacher, Madame
Basham, would tell of her travels, and of the
coffee shops and museums of Paris. When
my first daughter, Lainie, was born, I read
to her all the books about Madeline, the
little girl who lived with a nun in a boarding-

house in Paris. As Lainie grew up, she too began to have a fascination with anything involved with Paris or France.

Among some children's first visual references might be golden arches, causing them to point and scream, "Stop, stop, stop," when spotting a McDonald's, but my Lainie could spot a fleur-de-lis or anything French a mile away. If you think that reading to your children does not affect them, I have two stories that will prove you wrong. They still make me laugh and bring a tear to my eye.

The first story took place when Lainie was three. I'd discovered I was to have another child. At the time, Lainie was the center of Jay's and my universe. I knew she probably wasn't going to be too happy about sharing her life with a new baby. I tackled the situation head-on when I announced to her, "Lainie, we're going to have a new baby in the family and I want you to help Mommy and Daddy name it."

Lainie immediately blurted back, "Okay, we'll name the baby Idiot."

"No, Lainie, that is not a good name for a boy or a girl. Let's think of something special."

"Okay, then," she said without missing a beat. "If I can't name the baby Idiot — "

"No, Lainie, that would not be very nice. Let's pick a nice name, shall we?"

"Okay, then if it's a girl, we'll name her Madeline. If it's a boy, we'll name him Pepito!"

> Learn from the past, keep an eye on the future, but enjoy the present, for that is exactly what it is: a present. Life is all about the ride.

Pepito? Good Lord, what had I done? I loved the name Madeline, as we had read all the Madeline books together. But Pepito? Pepito was the name of the little boy who lived next door to Madeline, the Spanish ambassador's son. You can imagine my concern with a name like Pepito. Let's see, Pepito Patrick, little P.P. No, no, no, this would not do. Thinking on my feet, I quickly said, "I love the name Madeline, sweetie. On second thought, maybe we should let Daddy name the baby if it's a boy."

Lainie's first word was "lightbulb," which she had said as she pointed up at the ceiling while I changed her diapers. She had been speaking practically in complete sentences ever since. She nonchalantly replied, "Okay, Momma. Daddy can name it if it's a boy."

Then she went back to her task at hand — drawing an Eiffel Tower.

Thankfully, we had a little girl. Of course we named her Madeleine, spelled the same way author Ludwig Bemelmans's wife's name was spelled. I believe the right name for a person is just like the perfect title for a book, chosen only after great thought and consideration. I wanted my girls to have unique names, ones that would inspire them to live their lives like a great book: with passion, intrigue, great truth, and beauty.

When Lainie was very little she would proclaim, "When I grow up I am going to France to become a working artist."

I was especially pleased that she had added the "working" part, and when I inquired about that, she replied, "Momma, that's an artist who gets paid as opposed to one who paints for free." Hmmm, I liked the working part even better and the fact that at age three she had used the word *opposed.* Later, she decided that, besides being an artist, she would become a world-renowned chef (as opposed to a plain old chef). I am sure that seed was planted after she saw the video *Madeline Goes to Cooking School* and by the fact that I had conducted a Madeline Cooking School with chef William Stewart of the Stillwater Inn in Jefferson and his sous-chefs, Walter and Mallory, at the independent bookstore where I worked at the time. Again, all I knew was that I liked the *world-renowned* part.

I believe that in order for children to fly, they must first learn to spread their wings. I was reminded of the nanny Mary Poppins in the story by P. L. Travers. A stern disciplinarian, she still made learning fun, from magical housekeeping episodes to outdoor adventures in the park. I too would let our daughters experience all there was in life that we had to offer. I let them try everything within reason, things that they have an interest in while they are young. I think that, as we get older, we become less brave about trying new things. Children follow by example, and I have set the example of trying new things. I have found that I have enjoyed the musicals and plays I've taken them to just as much as the kids. After all, a spoonful of sugar does help the medicine go down.

Today Laynii (new spelling — she changes it every year) is thinking that maybe she would like to become the United States ambassador to France or a doctor. Or then again, after college, maybe she'll join the Peace Corps. Why not? I have always remembered what Dr. Joe Lester, one of the first authors to grace the doors of Beauty and the Book, told me. His momma used to tell him and his ten siblings, "If you can dream it, you can achieve it."

> Don't be afraid to dream. The only difference between a dream and reality is your state of mind.

Joe wrote a book called *I'm Not Afraid to Dream.* The book tells his story of how he and his brothers and sisters, the sons and daughters of a Georgia sharecropper and cafeteria worker, all rose to graduate from high school and college — and how a few went on for their master's degrees, two became dentists, and one went on for a Ph.D. Last I heard, Joe was running for Georgia state senate, and now I often give *I'm Not Afraid to Dream* to my friends and family as a graduation gift.

I will never forget what he told me: "We were so poor, my brother and I used to share a pair of underwear. Momma would wash it out each evening and we would take turns wearing it to school."

After he came to my shop, Joe told me that, being African American, he had never spoken at a general bookstore before, only at African American stores. He was surprised that my customers would be interested in hearing him. "Why would you think that, Joe?" I asked him.

He told me he figured that only the black-community-bookstore customers would be interested in buying his book. Joe also told me that he sold more books at the one event he had done in my shop than on most of his book tour. I told him that his book's theme was universal to all people. In fact, I thought he was doing a great disservice to book communities and bookstores by putting himself in a box.

Two brothers, Franz and Kurt Wisner, recently spoke at my shop. One had written *Honeymoon with My Brother,* which began with him getting dumped right before his wedding. With the wedding and honeymoon all paid for, he ended up having a big party and then asked his newly divorced brother, Kurt, to go with him on the honeymoon trip. Two years and fifty-three countries later, they came back home and Franz Wisner wrote the tale. He was talking to my book club about his book and telling us that everyone wasn't quite sure where it should be shelved: Nonfiction, Memoir, Inspirational, Travel. He said he couldn't believe it when he found out one mega-chain had actually shelved his book in Self-Help between Dr. Phil and Dr. Ruth.

When you try to label anything, you are limiting possibilities. We need to think more about perceiving things in the big picture instead of compartmentalizing everything into neat little boxes.

Whether you are raised in the rural farmlands of Georgia or the flint hills of Kansas or the coast of California, one thing is for sure: Through hard work and determination, you can achieve your goals. We do not have a choice of who our parents are or where we are raised. We *do* have a choice in what goes into our character and our spirit. No matter where I go in life, I will always be a small-town Kansas girl. I am very proud of where I come from and who my people are. I learned a lot from Joe and his book. I hope he learned something from his visit to my shop, too. If you have a good sense of place, sturdy roots, and a strong foundation, you can bloom to become whatever you want to be, no matter who you are or where you are from.

Yet while it is important to have roots, it is also important to be able to fly. What child hasn't dreamed of opening an umbrella and jumping out the window to fly as

Mary Poppins did? I know my own grand-father Dirt told me that, as a child, he too had jumped out of the hay barn holding an old black umbrella of his mother's and had just about broken both his legs.

I grew up with the song "Home on the Range." We sang it in music class at school, and it's as much a part of the Kansas landscape as "The Yellow Rose of Texas" is here in Texas. But as much as I love my home state and the great state of Texas, I always dreamed of travel. I think I must have sung "Supercalifragilisticexpiali-docious" as many times as "Home on the Range." I may have my little nest, but this bird has always been ready to fly.

One day I got a phone call from my photographer friend Patricia Richards, ask-ing if I would like to take my Pulpwood Queens Book Club members to Europe.

> "The right time is *any* time that one is so lucky to have."
> — *Henry James*

"Are you kidding, would I?" I said.

My bag was practically packed before she hung up the phone. Together, Patricia and I planned our first literary tour together. We called the trip the Pulpwood Queens Do Europe Literary Tour. Patricia had mapped out an itinerary where we would spend one week in Italy exploring the places our favorite authors had frequented, and the second week in France, where we would do the same. When everything was a go, I realized that there was no way I was going to go to France without Lainie: she'd named her sister after the Madeline books. She lived and dreamed of France, and she was just about to graduate from the eighth grade. I sat her down and told her that, if she would help me earn her passage, I would take her with me. Then I sat down with Madeleine and promised her that when she graduated from the eighth grade, I would take her on a similar trip.

Lainie worked her little tail off to earn the money to go to Europe. We applied for our passports and we pooled all our money, down to nickels, dimes, and pennies, so we could go. We were going to Europe!

We flew from Dallas to New York City, then on to Venice, Italy. Our days were filled with wonder as we walked the cobblestoned streets and rode in the gondolas of Venice. Our eyes were opened as we stared in awe at Michelangelo's *David* in Florence. We were fortunate enough to stay with the documentary filmmaker Frank Long, who had married into an Italian family that had quite a spread, shades of the television show *Dynasty* meets Tuscany. Frank arranged for our first international book signing that was held at his Villa Rucheli with author Sirpa Salenius and her book *Set in Stone: 19th-Century American Authors in Florence.* We later walked the walks of Oscar Wilde and Samuel Clemens, our American writer Mark Twain. We toasted with Bellinis in Harry's Bar in Venice just as Ernest Hemingway did when he ruled his domain there during his Italian days. We were welcomed with open arms as we spent time in Tuscany and partook of authentic Italian dishes made by the two Italian sisters who cooked especially for us each evening we were at the Villa Rucheli. The world history of my youth and days of armchair travel had finally come into being; I was on a grand tour with my daughter and the Pulpwood Queens.

The second week, we took water taxis from Venice to the train station. I hung on for dear life as we veered away from our moorings, and a man holding the same bar next to me inquired, "Where are you from in America?"

I told him, "I come from a small town in

427

> Go see the world now. What are you waiting for? If you linger for the time or the money, you will venture no farther than driving distance. It is true what they say: "If there is a will, there is a way."

Texas called Jefferson."

"Is the steam train still running in Jefferson?"

I was amazed he knew of our train and quickly responded, "Why yes it is. How did you know about our train?"

"Let me introduce myself," he said. His name was Martin. "I was the man who brought the train to Jefferson."

I could not believe my ears. Here I was, halfway around the world, yet I had met a person who loved my hometown as much as I do. We chatted all the way to the train station.

We boarded the night train and slept — or tried to sleep — as we traveled on to Paris. We had three bunk beds stacked on one side of the train car and three on the other, with barely enough room to turn around in the room. At least we could stretch out and sleep. Then I heard a bunch of girls laughing and carrying on down the outside corridor.

These girls came in and knocked on our open door and said, "Hey, y'all. We heard you-all talking from our sleeping berths down the way — are you-all from Texas?"

We busted out laughing as we learned that these girls were on a jaunt to Europe from Houston. Months later, they came by to visit me at my shop. Yes, it's a small, small world.

When we arrived in Paris, we picked up our rental vans from the train station. Let me tell you this, vans in France are not like the vans in Texas. They were squenchy. Some of us had to ride either sitting on luggage or with luggage on our laps. We didn't mind; we were excitedly anticipating our first view of the Eiffel Tower. I watched Lainie searching longingly out her window for the French landmark as we drove down the Parisian streets, and I saw tears start streaming down her face. Her eyes had caught the first glimpse of that architectural wonder. I began to cry, too, while Lainie exclaimed, "It's so beautiful, Momma, it's so beautiful," as the Eiffel Tower came into full view.

I didn't need to see it; I was seeing it through my daughter's eyes. I was crying just as much as she was. I was so choked with emotion I was afraid to speak. I will never forget that day, that moment, or that second.

Books do take you to magical places, from my little ole "Home on the Range" to the streets of Paris, France, and the magnificence of the Eiffel Tower. I often think back to my dear French teacher, Madame Basham. From the time I was Lainie's age until I graduated from high school, Madame Basham wholeheartedly and enthusiastically tried to teach us small-town Kansas kids "Bonjour, Madame" and "Au revoir, Madame" at Eureka High School. For years, she had led our French Club on trips to Kansas City to visit museums and eat in French restaurants. Growing up, I had never been outside of Eureka, Kansas — or the county, for that matter. I had traveled the world only from reading books.

These experiences opened our small-town Kansas eyes to the world of travel and the arts, just like my imagination had when I read the pages of a book. We just loved to get Madame Basham talking about the coffee, the café au lait; her eyes would just sparkle and shine as she gushed, "The coffee was so strong, mes petits chéris, that I would order a cup of espresso and a pot of hot water. I would pour the espresso into the pot of hot water and only then would I be able to enjoy my coffee."

In between readings of *The Little Prince,* Madame Basham would tell us about her shoes.

"Young ladies, I buy all my shoes in Paris, as they are exquisitely made. The French and Italians take great pride in their craftsmanship, and it is unparalleled."

I dreamed of buying shoes in Paris. I especially admired Madame Basham's navy blue calfskin pumps with little round circles on the top of the toes.

"Sometimes opportunities float right past your nose. Work hard, apply yourself, and be ready. When an opportunity comes, you grab it."
— *Julie Andrews*

We never got to go on a trip to France with Madame Basham. Even though she took a trip with her students every four years, no mention was made of us going my senior year. Somehow we knew not to ask. If I recall correctly, at that time she had been diagnosed with cancer. And while I never got to go on a French Club trip to France and Madame Basham has been gone for many, many years, I think she would be tickled pink to know I took my daughter on that trip years later.

In Paris, Lainie and I sat at a little street café and ordered our first café au lait. As we both sipped the foamy, rich, milky coffee, we took in the sights and sounds of the beautiful and historic city. I thought of Madame Basham and figured that she would be smiling down on Mademoiselle Virginie, as that was what I was called in class, and my Lainie. I can't think of anything that would have made her happier than passing on her love of reading, her stories, and her passion for travel to my daughter. Our last day of our European trip was in the City of Light. We picnicked at the park at the entrance to the Eiffel tower and marveled as the lights turned on this majestic architectural feat as the darkness dropped like a curtain. My arms wrapped around Lainie, we stared up into its wonder. Then I had an overwhelming feeling of homesickness for my Madeleine and Jay. Good-bye to Paris — it was time to go home.

What I learned from our trip was that, as much as I have loved being an armchair traveler, nothing beats the experience of traveling. What are you waiting for? The

> You will never understand how great home really is unless you travel to other places. And you'll never understand how great other places are if you only stay at home. Read a book, take a Sunday drive, plan a trip, and see the world. It's not only about the destination, but about the ride.

time? The money? The day may never come when we have time or money all at once. My advice to all of you is: Take the trip. If I could do it, anyone can. You may not fly first class or stay in five-star hotels, but it has been my experience that the people and the places will be the highlight of your trip, not your transportation or lodging. Just like in Michael Morris's novella, folks, "Live like you were dying." I plan on taking many more travel adventures, whether armchair travel with a book, road-tripping with my girlfriends, vacations with my family, or taking my Pulpwood Queens on literary tours abroad. In fact, we plan on going back to the Villa Rucheli in Italy. We will raise the funds by persevering and finding a way. All I know is when an opportunity arises, you must rise to the occasion. We'll call it the Pulpwood Queens Do Europe II and this time, besides the Queens, I am taking my fellow bookseller Fred McKenzie. I can't wait to see Italy through Fred's eyes.

Great reads that can inspire you on your life's travels, both near and far, include:

Jefferson: Riverport to the Southwest **by Dr. Fred Tarpley** *The* book on the history and folklore of my historic hometown, Jefferson, Texas. Special note: Available only through the Jefferson Historical Museum, which is the sole printer of this fantastic book on our "Belle of the Bayou."

***Texas Country Reporter* by Bob Phillips** The Charles Kuralt of Texas, Bob Phillips takes you to some of his favorite places to visit some of his favorite people off the beaten path.

***Ghosts of East Texas and Pineywoods* by Mitchel Whitington** The number-one best-selling book in my shop that tells the history, folklore, and legends of the ghosts and haunted places here in Jefferson and beyond in East Texas.

***Under the X in Texas* by Mike Renfro** This is one of my favorite read-aloud story collections, which I read in its entirety on a trip to Austin with my family for the Texas

Book Festival. I may have been hoarse by the time we got there, but we were primed and ready to hit one of the largest book festivals in the country.

***Close Calls* by Jan Reid** A contributing writer to *Texas Monthly* and other men's magazines, Jan writes in a way men will love — and women too. I call him a nonfiction Larry McMurtry.

***Cane River* by Lalita Tademy** Lalita Tademy decided to do a "Roots" story on her family and found out way more than she was expecting. Since the setting is just a couple hours from Jefferson, our book club visited Natchitoches, Louisiana, to take the river road to truly experience this tale of one woman tracing her family roots and turning it into fiction.

***Eccentric America* by Jan Friedman** I am featured in this book, so it is a must-read!

***Little Museums: Over 1,000 Small (and Not-So-Small) American Showplaces* by Lynne Arany and Archie Hobson** One spring break I took my girls, their cousins, and my husband's grandma on a trip that hit the off-the-beaten-path museums. On another spring break we went on an Elvis Presley tour through the South. You will never know who you are until you take some trips with your family.

The Little Prince by Antoine de Saint-Exupéry This is the first book assigned by my beloved Madame Basham to read in French in her class. I will never forget the adventures of the Little Prince.

The City of Falling Angels by John Berendt Our group recently read this book and now we are planning on going back to Venice, where this book is set, to visit the famous opera house, walk the cobblestoned paths, and traverse the canals of this city.

The Reluctant Tuscan by Phil Doran Hollywood's top television comedy writer Phil Doran's wife is an artist, and when visiting Italy she purchased a house in Tuscany. Phil Doran then went abroad to convince his wife that he wouldn't like it there, and he, too, fell in love with Tuscany. You will too when you read this book.

Set in Stone: 19th-Century American Authors in Florence by Sirpa Salenius In this book Sirpa Salenius focuses on stories and writings by the authors who have commemorative plaques in the city, including Mark Twain, James Fenimore Cooper, Nathaniel Hawthorne, Henry James, Henry Wadsworth Longfellow, and James Russell Lowell.

CHAPTER 16
MISS INTERNATIONAL

"I have ten commandments. The first nine
are Thou shalt not bore. The tenth is,
Thou shalt have right of final cut."
— *Billy Wilder, Austrian-born American
motion picture scenarist, director, and
producer*

Mrs. Perrier was my seventh-grade English
teacher. She was firm and ran the classroom
by the book. One day in class, a boy who
was already slipping into the category of
juvenile delinquent was asked by Mrs. Per-
rier to read aloud. (Half the time he was
playing hooky, and the other half he was in
class for attendance reasons only.) He obvi-
ously could not read even a basic sentence,
because each word he "read" was like pull-
ing teeth. Mrs. Perrier, ever calm, encour-
aged him, "Sound out each syllable, John.
Take your time and proceed."

As other boys started to snicker, all she

had to do was give them a look and immediately they became very serious and silent. I was painfully aware that John was in absolute misery. Mrs. Perrier always treated each student the same, so when it was John's turn to read, read he must. John would falter and hesitate with each word. Mrs. Perrier would quietly and patiently help him enunciate the word and then encourage him, "That's right, John, keep reading."

"You don't have to burn books to destroy a culture. You just have to get [people] to stop reading them."
— Ray Bradbury, American science fiction writer

I loved *The Adventures of Tom Sawyer* and *The Adventures of Huckleberry Finn.* Though I had not grown up on the Mississippi, I spent plenty of time in the creeks and on the banks of our nearby Fall River. I loved the stories of Tom and Finn and their outdoor adventures. As John faltered in his reading, I realized he was a boy like them, one who had a hard time sitting in school and was always daydreaming out the window. I thought to myself, If only someone read *The Adventures of Tom Sawyer* to him.

> "Today a reader, tomorrow a leader."
> — *Unknown*

John would end his paragraph and slump in his seat as the neighboring students gave him a high five. Mrs. Perrier would draw a bead on them and instantly they would sit straighter and give her their complete and rapt attention.

I really looked to Mrs. Perrier as a role model. She seemed taller than she actually was; she had perfect posture. Her voice was always level and well modulated. She never seemed to get angry or upset — she was always calm and serene. I found her fascinating. In my house everything was all noise and high drama. Since I was so shy and introverted, I liked her take-charge manner of leading the class and never getting exasperated, no matter the circumstances.

Her brown hair was an unobtrusive layered bob, not too big or too small. She wore clothes that were very demure, in monochrome hues of sage green, with straight skirts and sweater sets. Pearls and pumps completed her style of understated elegance and class. Always kind when asked a question, she never seemed to tire of any question thrown her way. With her quiet reserve, she made the class rise to the occasion, and she treated us with dignity and the utmost respect. We were, therefore, required to do the same. I have learned through the years that children do rise to the occasion when given the proper leadership.

> "The unread story is not a story, it is little black marks on wood pulp. The reader, reading it, makes it live: a live thing, a story."
> — *Ursula K. Le Guin, American writer best known for tales of science fiction*

One day Mrs. Perrier announced that there was going to be a seventh-grade spelling contest at Eureka Junior High School. All seventh-grade students were to be the initial contestants.

Mrs. Perrier explained that the winners of the in-class spelling contest would go on to compete in a schoolwide seventh-grade spelling bee contest held in the gymnasium. The winner would then go on to a state competition, and that winner would go on to the national spelling bee contest. As the grand finale, the national winner would go on to compete in the International Spelling Bee Competition.

I was an excellent speller. I had never missed a word on a spelling test. How hard could it be? I thought. I convinced myself that not only could I win the seventh-grade spelling bee, I could win the schoolwide bee at Eureka Junior High. Then I would go on to state and win the state spelling bee, and then to the Seventh-Grade National Spelling Bee, which, of course, I would also win. My final glory would be the ultimate, winning the seventh-grade International Spelling Bee.

"I have learned to spell hors d'oeuvres, which still grates on some people's n'oeuvres."
— *Warren Knox*

I could already imagine the fame and glory as thousands and thousands of people applauded my excellent spelling capabilities. I almost missed Mrs. Perrier giving out our homework assignment so lost was I in the daydream of finally getting my day in the sun, Miss International Spelling Bee Champion. Would they give me a crown or a tiara, a magical scepter or a dozen roses? I could see myself just like Miss America, walking down the ramp waving to all my admiring fans. Wouldn't my family be proud, shouting from the rooftops, "That's our daughter Kathy. Miss International!"

I quickly came back to reality when Mrs. Perrier said, "Here is your homework assignment and I will not repeat it again."

As I flipped open my notebook, I frantically jotted down the assignment.

I practically skipped home, reliving over and over the glory of receiving all the prizes and acknowledgment for my extreme spelling skills and intellect. Maybe then everybody would notice me and take me more seriously when I said I was going to do something. I might not have been good enough for a beauty pageant, but I could spell. I was certain I could win.

When I announced my intention to my family, they thought it was hysterical. Not only did they laugh their fool heads off, they nicknamed me "Miss International" and taunted me with it day and night. This was mainly from my mother and sisters, since my daddy was always gone. I doubt if he ever knew about their little nickname. It seemed that every time I mentioned anything about a test or a play tryout or running for a class or club officer position I was interested in, they would sing a strange little ditty they made up, set to the music of the International Coffee commercial: "Miss International!" with disparaging rhymes added. I would laugh and pretend it did not bother me, but it did — big time.

> "There are no secrets to success. It is the result of preparation, hard work, and learning from failure."
> — Colin Powell, former chairman of the U.S. Joint Chiefs of Staff and secretary of state

Secretly I was seething inside and thinking, I'll show them. Usually overwhelmed by their negativity, I escaped their taunting by finding a hiding place and reading. Like Tom Sawyer, I too was ready to run away. Oh, to float down the river on a raft — now that would be living. The place I chose to escape was usually under my bed or up in my tree house. I could hide from everybody and just let my mind take me to wherever the book I was reading at the moment was set. For a little while I could disappear into the pages with Tom and Becky. Their stories made me completely forget my own worries. I could ignore my family and continue with my latest attempt at triumph in its highest form. I would win the seventh-grade spelling bee.

Finally the day of the test arrived. I woke up that morning feeling totally pumped. Here it was, my big moment to shine. I dressed quickly in my favorite red, white, and blue outfit, and went down the stairs to gather my things for school.

"Hey, *Miss International,* this your *big* day?" my younger sister teased.

I tried to ignore her as I gathered my books and purse for school. She just kept singing "Miss International."

> Have faith in yourself. You are stronger than you think.

"Ha, ha, ha, very funny," I said. I tossed my long blond hair, the sides pulled back with a thick red yarn bow, sporting the latest trend, side corkscrew curls. I looked very much like Jan Brady from *The Brady Bunch,* which my sisters and I loved to watch and make fun of after school. Secretly, even though when together we trashed everything on that show, we all really wanted to be just like Marcia, the oldest, coolest sister. Her family lived in a really cool house and she had really nice, hip, cool parents. They never fought.

I slammed out the door and headed toward my best friend Heidi's house to walk to school. My mother rarely got up before we left for school; we girls were on our own in the mornings. We were expected to get ourselves up and ready and walk to school on time. I rarely had time for breakfast and even if I did, it would have been cold cereal with milk — something I could rustle up myself.

When my seventh-grade English and Literature period arrived, I marched into class confident that I was going to win, my head held high and adrenaline coursing through my veins. I greeted Mrs. Perrier and took my seat. I was ready.

I looked around the room, and all my classmates were just talking and kidding around and stuff. I could not understand why they were not taking this spelling bee more seriously. I arranged my notebook and pencils. The bell rang and everyone quieted down as Mrs. Perrier stood behind her desk and announced that we would begin the seventh-grade spelling bee.

"Students, please put away your books and supplies and have a clean desk."

> "Aristotle said that the essence of drama is fear and pity. Where do you feel more fear and pity than when you look at a kid in a spelling bee?"
> — *Rachel Sheinkin*

I placed everything on the shelf under my desk and felt my heart thumping in my chest. I was ready. Mrs. Perrier then said, "Class, everyone will participate in the spelling bee. I will announce your name alphabetically. You will stand to the side of your desk and I will read the word. If you like, I can repeat the word and use it in a sentence."

When my name was finally called, I stood up smiling broadly. Mrs. Perrier gave me my word: "Pneumonia."

Pneumonia? What kind of word was that? Everyone else had received easy words like *remember* or *brilliant,* and I got *pneumonia?* There was complete silence on my end as I racked my brain trying to think what letter *pneumonia* even began with. Sometimes my teachers would drive me crazy when I raised my hand asking them how to spell a word and they answered, "Kathy, you have a dictionary. Look up the correct spelling, please."

I mean, how could you look up the correct spelling if you weren't even sure what letter the word *began* with, for goodness' sakes?

I could not believe it. I did not know how to spell the word. I did not have a clue about how *pneumonia* was spelled. Sensing my dilemma, Mrs. Perrier repeated the word, pronouncing each syllable slowly: "Pneumonia."

I just stared at her blankly. My mind was going ninety miles an hour as I thought of every possible scenario for spelling the word. Nothing seemed even remotely right to me. I could not even figure out what the first letter was, let alone the rest. I was horrified that I could not come up with an answer.

"Would you like me to use your word in a sentence, Kathy?"

I shook my head no. I knew what the word meant. I just did not know how to spell it.

"Your time is up, Kathy. You may be seated."

Mrs. Perrier looked at me, surprised and dismayed. I had let her down. I dropped my head, afraid that — of all things — I was going to cry like a baby in front of the class.

How could this have happened? For the rest of the day a voice inside my head shouted, "You are an idiot." I did not dare look up at Mrs. Perrier or any of my classmates. I was sure they were going to say something to me or make fun of me in some way. I kept thinking, You are so stupid. You blew it.

I went straight home from school that day, running ahead of my friends so I would not have to face them. I was praying that my mother wouldn't be out of bed yet to ask me how things had gone. I was hoping that my sisters had forgotten and would be outside playing with the neighbor kids but this was not going to be my day. As I came in the door, there were my mother and sisters drinking cranberry juice in the kitchen. I went to get myself a drink. Before I could even get the cupboard door open, they began. "How did it go with the spelling bee, Miss International?"

As I placed the glass on the counter and opened the refrigerator to get the cranberry juice, I answered very seriously, "Not as expected, because I got this really hard word."

Before I could even finish, they all started laughing and said, "Miss International, don't tell us you misspelled the first word given!"

I poured the cranberry juice into a glass

and took a big gulp, trying to act as if it was no big deal. I started laughing and I began to choke. The juice went down the wrong pipe. I started coughing, and it was coming out my nose. The pink foam burned and my eyes watered as I tried to catch my breath. I couldn't breathe. Oh, the horror of dying a loser over some stupid cranberry juice.

> "I've come to believe that all my past failure and frustration were actually the foundation for the understandings that have created the new level of living that I now enjoy."
> — Anthony Robbins

I bent over the sink trying to get the juice out of my lungs, coughing and gasping for air. By now I was hacking and crying: I could not catch my breath. I thought, I am going to die. The cranberry juice just would not come out of my lungs. For some reason, with the pink froth from the juice coming out of my nose and all my coughing and sputtering, my mother and sisters all thought this was hilarious. They were bent over laughing and my sisters were falling to the floor in hysterics. I finally coughed so hard I got some oxygen in my lungs and gasped for air. As tears streamed down my face from the embarrassment of losing in the spelling bee and the humiliation of losing control of myself with the cranberry juice, I excused myself from the room. As I ran up the stairs, I thought, I'll show them. I will show them all someday. I will be somebody.

"If you ever start feeling like you have the goofiest, craziest, most dysfunctional family in the world, all you have to do is go to a state fair. Because five minutes at the fair, you'll be going, 'You know, we're all right. We are dang near royalty.' "
— *Jeff Foxworthy, American comedian*

As I flopped on my bed crying, I could hear them downstairs rehashing the whole scenario, laughing like hyenas. I instantly thought of poor Cinderella and the wicked stepmother and stepsisters. She got them back by marrying the handsome prince. I just wished somebody would come and rescue me. But no one ever did. I was on my own.

There were many, many more failed attempts at stardom. My senior year of high school, I convinced myself that I could become a cheerleader. Regardless of the fact that I am not in the least bit athletic and that I could barely do a somersault, I was bound and determined to give it a try. And a cartwheel? Oh, it was a wheel, all right: a wheel of misfortune. I can only imagine how ridiculous I must have looked executing that cheer. I had gotten a really popular senior cheerleader to help me. She looked so cool doing the cheer, but when I did it? What I lacked in talent, I certainly made up for in enthusiasm.

I see many kids today just like me, trying so hard to be accepted and to excel at something — anything — but failing. I will stop whatever I am doing and take the time to help them find out what they can do and do well. Thank God for all those mentors who helped guide me to the right path and gave me hope. I had many more next times and many more failures, but the important thing is that I never gave up and never stopped trying new things. I simply pulled myself right back up by the bootstraps and began again.

There is a wonderful expression I once heard, which is attributed to the Dalai Lama: "When you lose, do not lose the lesson." I did not win that seventh-grade spelling contest, but I did learn an important lesson. Everything I know I learned from my mistakes.

Only now, instead of mistakes I call them discoveries. In order to be the best that you can be, you must continue to educate yourself. You must be a lifelong learner. And as my good friend Fred McKenzie always says, "Once you stop, you drop."

> "Twenty years from now you will be more disappointed by the things you didn't do than by the ones you did do: So throw off the bowlines. Sail away from safe harbor. Catch the trade winds in your sails. Explore. Dream. Discover."
>
> — Mark Twain,
> American humorist, writer, lecturer

You may not have graduated from high school, but it is never too late. Go back to school — you can do it. If you graduated from high school, go to college, go on for your master's, your doctorate. Those are excellent choices, but to truly educate ourselves, we must be lifelong learners. We must walk in another person's shoes, we must learn of other people and cultures. I can think of only one way to do this: read. Read every day to educate, to enlighten, to broaden your world.

> Never think that an education makes you better than anyone else. You may have just had opportunities others did not. What truly makes one educated is in helping those up from behind you. Be a mentor.

You also must understand that not everybody has your best interests at heart. In order for you to succeed and be a complete person, you must learn that you are responsible for your own actions and for yourself. Do not let others stand in your way or knock down your dreams. Look for others who will encourage you and help you find your way. Focus on your priorities, develop an inner strength, and concentrate on what you want to accomplish.

You have heard that old saying "We are our own worst enemy." We certainly can be, but I say, "We need to be our own best friend."

When we stumble, we ask ourselves, "What if? What if I had done this? What if that had not happened?" Those are not the questions to ask. Instead, ask, "How can I learn from this experience and do better next time?"

In Pat Conroy's wonderful book *The Great Santini,* I learned that those you love the most can be your greatest enemies. I have come to learn that when others try to make you feel bad about yourself, they are really just unhappy with their own circumstances. We need to create our own happiness and not be dependent upon others for that accomplishment.

In June of 2002, the Pulpwood Queens were asked to kick off Diane Sawyer and Charlie Gibson's Read This! book club. Earlier that spring, we had been invited to appear on the show. I surmised that we had passed their muster and were on our way to promoting literacy. I think that this was one of the greatest experiences for the Pulpwood Queens: finally to have been given that kind of credibility on our mission.

My mother called me shortly after that show and, when I answered the phone, she never even said hello or told me who she was. She just began, "Well, I was really proud of you today."

I couldn't believe it. My mother was finally proud of something I had done, and

> "I think the one lesson I have learned is that there is no substitute for paying attention."
> — *Diane Sawyer,*
> *news anchor, reporter, and journalist*

it had only taken me about forty-some-odd years to accomplish that goal. I had mixed emotions about her statement.

She continued, "You know I never liked your name. Your father named you Kathy. I always wanted your name to be something like Angelique or Veronica. Well, anyway, when Diane Sawyer said 'Kaaawtheee Paaaaawtreek,' I thought that it was the most beautiful name I have ever heard."

My mother hated my name. I mumbled something like "That's nice, mother. I'm glad you now like my name."

Then she went off on her usual tangent. "It's kind of like when I was out in Hollywood . . ."

I tuned out after that, because all I could think was that my mother had just said she hated my name and was proud of me at the same time. It was funny, but this was something I had waited for my entire life.

My mother had just said she was proud of me.

As I hung up, I began laughing, then started laughing hysterically. I was bent over laughing and then I just sat up and stopped. I realized that the one thing I had been waiting for my whole life was something I'd had with me the whole time: a sense of self-worth. At that moment I also realized that your family is your family. They carry with them the baggage from their upbringing, and that heavy cargo often travels from one generation to the next. Instead of looking around for someone to blame, get your own house in order. When you move in, unpack only what you need and let the rest go. I believe that what hurts us most also makes us stronger, and the good Lord does not give us more than we can carry.

> "What we become depends on what we read after all the professors have finished with us. The greatest university of all is the collection of books."
> — *Thomas Carlyle, Scottish historian and essayist*

I love my family. I love them for all their imperfections. I also know of only one person who was perfect. I have found great solace and peace in the Good Book and in good books. They are my constant companions. I keep them close and near, and cherish the feel of the page, the smell of the ink, the power of the words.

All of us have our own experiences and upbringings, and we all have to find our own way to know who we are and why we are here. Sometimes things are beyond our control. I have always reached for the stars and, in that, I am not that much different from my mother. For myself, I have found that reading is the perfect ladder to get there. You may never actually touch the stars, but you just might reach the heavens. As I have often said, whether I succeed or fail, I would lie in my grave peacefully with just two words for my epitaph: "She tried."

Every day in my shop I am asked, "What would you personally recommend?"

My reply is usually, "Have you ever read any Pat Conroy?"

I am continually amazed that people will answer, "No. What are his books about?"

His books are about me, about you, about finding your place in this world. His stories saved me and told me I was not alone. I was not alone.

> "If I read a book and it makes my whole body so cold that no fire can ever warm me, I know that is poetry."
> — *Emily Dickinson, American poet*

So that is my mission in life: putting good books in the hands of readers. Because of Pat Conroy's books, I am writing this book today. His books changed my life, and I knew the minute I read them that I was going to be okay. We read so we can find ourselves. There is great power in the written word. It's the stories, it is all in the stories. I believe that I am becoming a writer so that others like me will not be alone. Every writer who visits me and my book clubs offers something to be learned. I have learned that writers are my people. We accept each other with all our flaws and imperfections. That is why it is so important for me to help authors who are talented and gifted get discovered. We are all looking to find our place in this world, and it is through the words of authors, poets, and — yes — even songwriters that I have found my place. I have built a house of books and now I am at home at last.

I will continue to read and reread everything Pat Conroy will ever write, along with the work of my other favorite authors, such as Ellen Gilchrist, Mark Childress, Doug Marlette, Cassandra King, Tennessee Williams, Harper Lee, and Michael Lee West. Yes, they are all Southern authors. Their stories speak to me in a way no others do. I find great peace in their prose, their language. You are how you are raised. I love Southern food the best, Southern music the best, Southern people the best, and Southern authors the best, but they are all my springboard toward discovering a big ole world out there, and I am happy to branch out, to travel through books to places I have never been and then eventually to go there. I want to see for myself, taste for myself, and feel for myself. Books have taken me this far and I have the feeling that, before my life is over, they will take me even farther. I am now ready for that journey. Are you?

I could not end without giving just a few more of my personal favorite reads:

The Bridge by **Doug Marlette** Pulitzer Prize winner Doug Marlette's first novel of family, forgiveness, strong women, and headstrong men, this is a love story from a gifted storyteller. I look forward to all of the books he writes.

Drunk With Love by Ellen Gilchrist Ellen Gilchrist is a creative genius when it comes to short stories. I have loved short stories since I discovered O. Henry in junior high, and for the best, look no farther than _Drunk With Love._

Cold Sassy Tree by Olive Ann Burns I was given this book to read when I first moved to Jefferson; I was told, "This book could have been set in Jefferson." They were right and you will fall in love with Olive Ann Burns's storytelling. Unfortunately, she died of cancer not too long after _Cold Sassy Tree_ was published.

The Last of the Southern Girls by Willie Morris When I read this book, I became the biggest Willie Morris fan who ever lived and read everything of his I could put my hands on. Though this story of a small-town Southern girl making good in our nation's capital was written some time ago, the story rings as true today as it did when it was written.

A False Sense of Well Being by Jeanne Braselton You will have to read this book to understand Jeanne's marvelous sense of humor. I had chills down my spine when I heard of her death. She was a wonderful writer, and she should be remembered.

***Touched* by Carolyn Haines** Carolyn is a storyteller supreme. *Touched* was one of the first books of hers I read, and I have been a big fan ever since. A mail-order bride marries the town barber and befriends a woman whose daughter gets struck by lightning, only to inherit the gift of being able to tell the future.

***Father and Son* by Larry Brown** One of the best writers I have ever read and nicest authors I have ever met. This story is about a man who returns to his Southern hometown and the five days that follow. A psychological thriller that will keep you turning the pages. You'll never forget the story or the author.

I cannot end this book without telling you that what I have written here is just my opinion and my take on life in general. I am and will always be a small-town Kansas girl. I encourage you to find your own guide to life by listing the books that are important to you, writing down the principles that are important to you, and telling your own story. Let these things be the compass to guide your life.

After writing this book I found that until I put my thoughts and beliefs on paper, I really did not even know what I truly believed in as a human being. I know now and I feel I have come full circle in my life. I am ready to do good works, continuing to promote literacy.

Someone once told me that every time a person dies we have lost a library of books filled with great stories. I hope that my book will be a springboard to helping you find your way on the rapturous road to reading and the wonderful world of books.

> "When a book leaves your hands, it belongs to God. He may use it to save a few souls or to try a few others, but I think that for the writer to worry is to take over God's business."
> — *Flannery O'Connor, American writer*

■ ■ ■ ■

RESOURCES

■ ■ ■ ■

THREE MORE SUGGESTED READING LISTS FOR AVID READERS

ANIMAL TALES

In grade school, right after lunch break, our teachers would read to us for half an hour. One or two chapters of a book would get us all settled back down from our recess and back into school mode. Our teachers read books like *Bristle Face, Old Yeller, Mr. Popper's Penguins,* and *Rascal.* We loved those stories. To this day I can remember exactly how choked up I felt when Laura Ingalls Wilder lost her dog Jack out on the prairie. When Mrs. Hall first read to us about Laura and Jack, all the children around me cried so hard that she had to pass out tissues. Those stories taught us about life, love, and true devotion. They also taught us that life is about change, and that it's sometimes very hard and unfair. What I remember the most is the undying love with a pet that always endures.

Some of my favorite stories today remain the animal tales read to me by my teachers after recess. Here they are, and a few more that I have read and shared with my daughters.

Summer of the Monkeys by Wilson Rawls
Bristle Face by Zachary Ball
Old Yeller and *Savage Sam* by Fred Gipson
Mr. Popper's Penguins by Richard and Florence Atwater
Rascal by Sterling North
Charlotte's Web by E. B. White
Black Beauty by Anna Sewell
My Dog Skip by Willie Morris
My Cat Spit McGee by Willie Morris
The Yearling by Marjorie Kinnan Rawlings
The Incredible Journey by Sheila Burnford
Susie, the Whispering Horse by Dr. Michael Johnson
"Canticles to a Road Cat" by Robert James Waller

THE ENVELOPE, PLEASE!

This past year I hosted an Academy Awards party at the Big Cypress Bayou Coffee Shop and Internet Café. Duane, the owner, had recently purchased a big-screen TV, and I jumped at the opportunity. I told him I would do all the decorations and award-winning snacks, and we would purchase our drinks from him. Duane gave us a big thumbs-up. Instead of a more traditional evening-gown-and-black-tie affair, we invited everyone to come in their pajamas!

All the Pulpwood Queens arrived, walked the red carpet, and were seated at their festive tables. As the show began, the Big Cypress was filled with opinions on the movies we'd seen and on Jon Stewart as the master of ceremonies, as well as screams of, "What was she thinking? Could she get a bigger bow on that shoulder?"

The evening progressed, and we cheered or moaned over whether our favorite movies had actually won. We all commented how almost all the nominated films had been based on a book. It was true. I guess it is easier to adapt a screenplay from a book than to write an original one.

I've always been passionate about books, but I also believe that sometimes there's no better entertainment than a good movie. Here are some of my favorite reads that were made into some of my favorite films. These are perfect for combined book-and-a-movie book-club discussions.

To Kill a Mockingbird by Harper Lee

Breakfast at Tiffany's by Truman Capote

Cold Comfort Farm by Stella Gibbons

Empire Falls by Richard Russo

The Color Purple by Alice Walker

Steel Magnolias by Robert Harding

"Rita Hayworth and Shawshank Redemption" from *Different Seasons* by Stephen King, made into the film *The Shawshank Redemption*

"The Body" from *Different Seasons* by Stephen King, made into the film *Stand By Me*

Rocket Boys by Homer Hickam, Jr., made into the film *October Sky*

Texasville by Larry McMurtry, made into the film *The Last Picture Show*

The Hotel New Hampshire by John Irving

Forrest Gump by Winston Groom

Freaky Friday by Mary Rodgers

Legally Blonde, based on characters created by Amanda Brown, story by Natalie Standiford

Fried Green Tomatoes at the Whistle Stop Cafe by Fannie Flagg

The Birds by Camille Paglia

"Brokeback Mountain" from *Close Range* by Annie Proulx, made into the film *Brokeback Mountain*

Memoirs of a Geisha by Arthur Golden

Often, authors who try to break into fiction after having already earned a name in another profession find a hard time with the critics. For whatever reason, they are not seen as credible or as "important" as authors who always worked as professional writers.

Linda Bloodworth Thomason, a television writer and film director, comes to mind. One of our top television writers, responsible for wonderful series like *Designing Women* and *Evening Shade,* writing a book? How could she?

Or author Doug Marlette, a Pulitzer Prize–winning political cartoonist, trying his hand at fiction? Or actress-turned-author Ronnie Claire Edwards, most famous as the character of Cora Beth Godsey on the show *The Waltons*? Of course, Thomason's *Liberating Paris* was one of the best books I have read in a long time.

I love reading books by authors who have taken up writing as a second career in life. These writers have broken out of a box, defied expectations of what they were supposed to do, and they have written some phenomenal stories. Those that I have found particularly memorable are:

Liberating Paris by Linda Bloodworth Thomason, creator of the television series *Designing Women*

The Scandalous Summer of Sissy LeBlanc by Loraine Despres, most famous for writing the episode "Who Shot J.R.?" on *Dallas*

The Bridge by Doug Marlette, winner of the Pulitzer Prize for his political cartoons

Beaches by Iris Rainer Dart, former comedy writer for Cher on *The Sonny and Cher Comedy Hour*

The Song Reader by Lisa Tucker, former jazz singer

Goodbye, Little Rock and Roller by Marshall Chapman, musical artist and songwriter

The Knife Thrower's Assistant by Ronnie Claire Edwards, actress and playwright

Beyond the Blonde by Kathleen Flynn-Hui, haircolorist

HOW TO START A BOOK CLUB

Just as there is more than one way to skin a cat, there are different ways to set up a book-reading group. Many groups are small, made up of six to eight friends and acquaintances within a community who get together to read and talk about books simply because they like to read. They might meet once a month at each other's houses and decide on books to read by consensus — someone will suggest a title, and if it sounds good to everyone, that will be the pick. Or they may take turns making the selection and everyone agrees to go along. Publishers have long been aware of the popularity of book clubs, and have for some time printed up reading-group guides, available for free in bookstores, consisting of an interview with the author and sample discussion questions that can help get the ball rolling on meeting nights. A few even bind these guides into the back of the paperback edition of the book.

Today there are even more types of clubs, including on-line book clubs and television book clubs, and all of those clubs are great. But if you've come this far in my book, you already know that we Pulpwood Queens do things a little differently. Ever since the day my sister Karen put the beauty shop/bookstore bee in my bonnet, I knew I wanted authors involved in some way. In my years of working in books on all sides of the bookselling tables, I have learned how to go about getting authors actually to come to the meetings, and that has made the experience so much more exciting for my Queens. Maybe some of you out there would like to start a Pulpwood Queens group or join an existing chapter in one of many towns and cities across the country. Or maybe some of you want to start your own book club, in which case I say "Hallelujah." There are plenty of good books to go around. A publisher once printed some promotional T-shirts for the annual booksellers convention that read SO MANY BOOKS, SO LITTLE TIME. So true!

And if you want to join the largest "meeting and discussing" book club in the cosmos, the Pulpwood Queens, contact me at

Kathy@beautyandthebook.com.

Here is a Pulpwood Queen's considered advice on what works and what doesn't:

1. Don't limit membership. Call all your friends and ask them to join. They don't have to be readers. Some of my most enthusiastic members weren't big readers before they joined the group. They came because it seemed like fun, and isn't that the way reading is supposed to be? Unlike those who prefer smaller, more homogeneous groups, I say, the more the merrier. My larger chapters seem to ebb and flow as members move away and others join the fun. I think discussions are livelier with a diverse group, and besides, if my members don't know each other well before they join, they tend to become friends quickly. You can learn a lot about someone after you have been talking about a book for a couple of hours. It is absolutely mandatory for all members to read the book. I know we all have busy lives. There will always be times when you just can't get the book finished in time. But I feel that, to truly experience a book-club meeting (especially when the author is in attendance either in person or by phone), reading the book is common courtesy. I don't make coming to the meetings mandatory, and my Pulpwood Queens appreciate the flexibility. They don't have to worry that they will let the group down by missing a meeting. Even though not every member comes to every meeting, each chapter has a solid hard-core group. Now, you may be saying to yourself: Well, that's fine, but what if *too* many people show up? Honey, all I can say is I once fit over 200 people in my house for an author event. We were packed in like sardines, but everybody had a ball!

2. Choose a leader, someone who is committed to running the meetings and keeping the discussion on track. Ideally, she will be an upbeat, enthusiastic "people person" who can lead the group without being either uppity or condescending or lording it over the group in any way. Each Pulpwood chapter has a Head Queen. I love being Head Queen and so does each of my Head Queens and Head Timber Guys (men who run chapters of my book clubs). Being an avid reader and not at all shy about voicing my opinions, I feel I have received the calling in the way that others receive the calling to become ministers, teachers, or doctors.

3. Choose your location. You can take turns meeting in each other's homes, or one or two members with the most room and fewest potential distractions, like noisy, nosey, "Hey, can I come, too, Momma?" kids, might rotate hosting duties. Many local bookstores are happy to play host to reading groups, as are libraries. Be creative. Some Pulpwood Queens chapters meet in bed-and-breakfasts or college coffeehouses. I even have one chapter that meets in the Texas Rock & Roll Hall of Fame, Music City Texas!

4. Choose a day, date, and time to meet and try to make it consistent from month to month. My Pulpwood Queens meet on the first Tuesday of every month. Once in awhile, we will move the meeting due to a holiday or a guest author's schedule. Most of the time we try to keep some sort of consistency. That way nobody can say they never can remember when the book club meets.

5. Prepare discussion questions before the meeting. This is the leader's responsibility, but there is no reason for anyone to feel intimidated by this task. I provide the questions for the Pulpwood Queens Book Club chapters on our Web site, but you can also e-mail a book's publisher or the author, and they are usually happy to send you questions. There are also a number of Internet sources and sites that suggest general questions. Each of us responds slightly differently to the characters and events in a novel, as we all bring our own experiences into how we interpret the meaning of a book. Just remember that there are no bad questions and no wrong answers. I often find that the part of a book that moved me the most didn't move other club members as much, and vice versa. And that is just how it should be — like life. I believe that our personal experience of a book, the way we relate to or identify with each character, is what reading is all about. It's what stays with us long after we have forgotten plot details or minor characters. Of course, I'm sometimes amazed at comments I never saw coming. I once had a member object to a book because she said the main character, a Methodist minister's wife, "would never act that way." She got quite excited about it, and I had to remind her that the book was *written* by a former Methodist minister's wife. So again, there are no wrong answers, but discussions can get lively. Sometimes I like it when there is disagreement over a book, because then the conversation goes on even longer.

Six Questions Guaranteed to Get a Discussion Going

1. At what point did you really become involved in the story?
2. What character did you relate to most?
3. What do you think is the author's underlying theme?
4. Did you feel that the book was believable?
5. If this were going to be a film, whom would you cast as the main characters?
6. If you could change one thing in this book, what would it be?

What Not to Do: Four Ways to Absolutely Kill a Conversation

1. Allow one member to hog the discussion. There is usually someone who loves to pontificate. Some of my Pulpwood Queens might be thinking that person is me. Guilty as charged. I do love to talk books. Feel free to give me a little nudge and shut me up.
2. Start with the question "Who loved this book?" If everyone raises her hand, nothing more is left to be said.
3. Ask questions that can be answered with a yes or no.
4. Select a book that has excessive profan-

ity. Since most of my book-club members are from the South, they tend to get very upset by foul language. I also happen to believe that, unless profanities contribute to a mood or setting in the story, there are other words that would be more appropriate.

WAYS TO TAKE YOUR LOVE OF BOOKS TO YOUR COMMUNITY

1. Donate new or used books to your local town or school libraries. In my local Jefferson chapter, we have a program where, on her birthday, each Pulpwood Queen donates that selection of the month to the Jefferson Carnegie Library. I also ask each visiting author to donate a copy of her or his book to our libraries or schools. Most are more than willing to do so.
2. Volunteer at your local library, schools, nursing homes, and assisted-living facilities to take part in read-aloud story hours.
3. Work with parents' associations in your local communities to sponsor used-book sales in order to raise money to support your local libraries and schools.
4. Adopt Seattle librarian Nancy Pearl's citywide reading program, One Book, One City. The Pulpwood Queens partnered

with the *Marshall News Messenger* and went one step farther by getting volunteers to read the entire text of *To Kill a Mockingbird* at a very public place to promote literacy.

5. Partner with other organizations on literacy efforts. The Jefferson Rotary Club and the Pulpwood Queens did just that to bring an award-winning children's author to our schools to do writing workshops for high-school seniors.

Steps to Making Reading a Family Affair

1. Read, read, read to your kids beginning at an early age, and don't stop until they can take over for themselves. Hint: This is later than you think. Younger kids who are learning to read often enjoy listening to you read more advanced chapter books. Better yet, do a lot of reading yourself. Children learn by example.

2. Use your local library. Get your child a library card and encourage its use. Make one day of the week Library Day. Many libraries sponsor reading programs for young children. Sign them up and stay to listen yourself. Show your kids that you are interested, too.

3. Don't censor kids' reading choices. Many a writer and filmmaker started out with *Mad* magazine and comic books. Teach them to read for pleasure and entertainment, and you will have a lifelong reader.

4. Limit TV time. Experts can debate the harm TV does or doesn't do 'til the cows come home, but the fact is that an hour in front of the TV is an hour a child isn't reading. In our sped-up, overscheduled, media'd-to-death age, it's up to us adults to give kids the *opportunity* to read. That means giving them the *time* to read.

THE PULPWOOD QUEENS' FAVORITE RECIPES

APPETIZERS

It's a Mystery to Me Appetizers

I first heard of a rendition of this recipe when I was given an electric kabob cooker for Christmas. I made it my own with peppered bacon and brown sugar. No man can resist these little doobies, and the girls love them too.

Ingredients:
1 package peppered bacon strips
1 large can chunk pineapple, drain juice
1 large jar pitted green olives, the big ones
1 package pitted prunes
1 cup brown sugar
Toothpicks, the fancy ones with ruffled edges

Cut the bacon in half lengthwise so you have two equal blocks of strips. Take one strip and, starting with the little end, wrap around a hunk of pineapple, an olive, or a prune and secure with a toothpick. Roll each one gently in a bowl of brown sugar. Place appetizers on a Pam-sprayed baking sheet, being careful not to let them touch. Bake at 350 degrees for 15 to 20 minutes. Serve hot.

I don't tell anybody what is inside the appetizers, thus the mystery. When they get the prune, they never guess in a million years what it is. *Dee*-lish!

Marla-rita's Cheese Ball and Crackers

I can make a meal of Marla's cheese ball and crackers. Served at all our Hair Balls, the following recipes are courtesy of my co-hairstylist Marla Keith, my favorite cook in the whole wide world! There is nothing Marla makes that isn't to die for. Now you take a woman who can cook, clean, do hair, take incredibly flattering photographs (she took my book-jacket photo), who is smart, funny, and the best friend to me, and, girlfriends, you have someone rare and fine. Besides, she's knockdown-dead gorgeous and also can whip up a margarita better than anybody! That is why we call her Marla-rita! Her cheese ball and crackers are a perfect match.

Cheese Ball

Ingredients:
1 brick sharp Cheddar cheese, shredded
1/2 cup finely chopped yellow sweet onion
2/3 cup Hellmann's Real Mayonnaise
Pinch of garlic
1/4 teaspoon garlic powder
1/2 teaspoon cayenne pepper
1 1/2 to 2 tablespoons cracked black pepper
1 cup halved pecans, toasted and cooled

Mix Cheddar cheese and onions in a bowl. In a separate bowl, mix most of mayonnaise, garlic, garlic powder, cayenne pepper, and cracked black pepper. Add more mayonnaise until just moist. Add mixture to cheese and onion until just moist, then add toasted pecans. Shape into a ball and refrigerate until time to serve.

Crackers

Ingredients:
1 1/4 cup canola oil
1 package Original Ranch dressing mix
2 tablespoons crushed red pepper
1 box Zesta saltine crackers

Mix oil with Ranch dressing mix and crushed red pepper. Get a large plastic container with lid and layer in crackers, drizzle on the liquid spice ingredients, another layer of crackers, another drizzle, crackers, drizzle. Continue until all crackers are coated with the liquid spice ingredients. Pat gently to make sure all crackers are coated with mixture. Set aside for twenty-four hours before serving.

Man may not be able to live on bread alone, but here is one woman who certainly could live on Marla-rita's Cheese and Crackers.

Thanksgiving Meatballs

Bill Reed, husband of our charter chapter's Hospitality Hostess Dona Reed (and pastry chef and caterer extraordinaire) makes these. All I have to say is "Gobble, gobble!"

Ingredients:
1 pound sausage
1 6-ounce package Stove Top Stuffing Mix for chicken
3/4 cup whole-berry cranberry sauce
1 egg
1 cup water
1/4 cup melted butter

Preheat oven to 325 degrees. Brown sausage and drain. In a large bowl, combine sausage and stuffing mix. Stir in cranberry sauce, egg, and water. Shape mixture into

16 balls. Place balls on a foil-covered baking sheet, and brush balls with melted butter. Bake 20 minutes. Preparation time: 30 minutes.

Bill says: Number of servings: 8 (2 balls each), but remember the Queens leave our "diets at the door." I'd double that recipe.

SALADS

Wanda Yates's Coleslaw

Our Beauty and the Book clients often give us great recipes. Here is one recipe from a favorite client we just adore. We often serve these at book-club meetings.

Ingredients:
16-ounce package coleslaw
1 bunch green onions (sliced with tops)
2 packages beef ramen noodles (save flavor packages for dressing, below)
1 cup toasted almonds
1 cup sunflower seeds

Dressing Ingredients:
Ramen flavor packages
1/2 cup sugar
3/4 cup oil
1/3 cup white vinegar

In a large bowl, combine the coleslaw and

green onions. In a separate bowl, combine the ingredients for the dressing. Pour the dressing over the coleslaw and onions. Refrigerate for at least 4 hours. Immediately before serving, add the noodles, almonds, and seeds.

Grandma Murphy's Green Salad

This recipe comes from my daddy's side of the family and was always served at Thanksgiving and Christmas as a side dish for the turkey. Very tart and tangy — one bite reminds me of being home for the holidays.

Ingredients:
1 cup chopped celery
1/2 cup sweet pickle relish
1/2 cup chopped green onion
1/2 cup chopped green olives
1 cup chopped head lettuce
1 large box lime Jell-O

Place all chopped ingredients into a Jell-O mold or 9-by-13-inch glass dish. Prepare Jell-O according to directions on box and pour over chopped ingredients. Chill 'til set.

Southern Pulpwood Queens Pecan Trifle

I can make this dish with a bowl and a spoon on my manicure table right before book club in about five minutes.

Ingredients:
1 store-bought pecan pie (I buy mine at the local Brookshire's grocery store)
1 large container whipped topping
1 Heath bar, crumbled
A really cool glass footed trifle dish

Tump (tip out gently) the pecan pie into a large mixing bowl, break and chop up with a large spoon into bite-sized pieces. Plop in whipped topping and stir until evenly distributed among the bite-sized pie pieces. The mixture will be light brown and lumpy. Spoon it delicately into the glass trifle bowl, then top with crumbled Heath bar.

When everyone oohs and ahhs, just say, "Oh, darlings, it's an old Southern family recipe," and wink. Besides, it's my youngest daughter's favorite dessert; it's easy and quick, and it leaves way more time for reading.

Peanut Butter Fudge Cake

Jean Wright of the Pulpwood Queens of Linden, Texas, served this at a book-club meeting and I am telling you, Elvis would have loved this stuff!

Cake Ingredients:
2 cups sugar
2 cups all-purpose flour
1 teaspoon baking soda
1/2 teaspoon salt
1 1/2 cups creamy peanut butter
1 (8-ounce) container sour cream
2 large eggs, beaten lightly
1 cup butter
1 cup milk
1/4 cup cocoa

Kat's Coffee Fix Fudge Frosting
Frosting Ingredients:
1/2 cup butter
1/2 cup filled with milk and a shot of espresso coffee, but make sure it's only 1/2 cup
1/4 cup cocoa
1–1 1/2 (1-pound) packages powdered sugar
1 teaspoon vanilla extract

To make the cake, combine sugar, flour, baking soda, and salt in a large bowl; stir in peanut butter, sour cream, and eggs. Melt the butter in a saucepan over low to medium heat. Whisk in the milk and cocoa. Mix with a beater on low until thoroughly blended. Pour into a greased 15-by-10-inch jellyroll pan. Bake at 325 degrees for 25 to 30 minutes or until a wooden toothpick inserted into the center of the cake comes out clean. Spread the frosting over the warm cake. Frost with Kat's Coffee Fix Fudge Frosting.

To make the frosting, melt the butter in a saucepan over medium heat. Whisk in the espresso with milk and cocoa; bring mixture to a boil. Remove from heat. Gradually add sugar, stirring or using an electric mixer on low until smooth and the consistency you want; stir in vanilla. Cake may be garnished with peanut-butter-cup-candyhalves, if desired. If icing becomes too thick, thin by adding 1 to 2 tablespoons of coffee.

Great-Grandma Chisholm's Green Tomato Pie

Who would ever think that a pie made with tomatoes could be as rich (and sweet) as any you have ever tasted? This recipe comes from my mother's side of the family and has to be tasted to be believed. It tastes just like apple pie, except way richer. This is what they used to make when apples weren't readily available and too expensive to buy. The pie is fantastic with homemade vanilla ice cream but that's my daddy's recipe, and he hasn't given it to me yet.

Ingredients:
1 cup sugar
1 teaspoon cinnamon
6 large green tomatoes or 8 small (they must be totally green tomatoes with absolutely no signs of ripeness or redness), sliced
Juice of one lemon
2 tablespoons butter
Pie crust for top and bottom — use your favorite recipe

Divide dough into two equal balls. Roll out dough for upper and bottom crust. Grandma Chisholm, who personally taught me this recipe, once said to me, "Now sister, work that dough as little as possible to get the flakiest crust and don't use too much flour when you roll it out. It makes the dough tougher than that old rooster we cooked last Sunday." Combine the sugar and cinnamon and set aside. Place bottom crust in pie pan and as you layer in the tomatoes, sprinkle with cinnamon sugar.

Pour the last of the sugar cinnamon on top, squeeze on lemon juice, and dab with butter. Top with upper crust. With a knife, trim excess dough and pinch into flutes around the top of the pie. All four-foot-nine of my grandma Chisholm used to do this lickety-split. Slice a design in the top center to let steam escape. She used to slice KATSOUP into the top as that was my nickname (based on my mispronouncing "catsup" when I was first learning how to read). Bake on a cookie sheet for 1 hour at 350 degrees and check the pie as it bakes. If the top looks too brown, cover the edges with aluminum foil.

My Truly Scrumptious Chocolate Present
Here's another recipe that will leave them thinking you are cooler than Julia Child and the Naked Chef all rolled into one!

Ingredients:
1 sheet of boxed Pepperidge Farm frozen pastry, thawed
1 package of snack-size Snickers bars, unwrapped
1/4 cup melted butter

Place pastry sheet on a cookie sheet sprayed with Pam cooking spray. Pile Snickers in a mountain in the center of the pastry sheet. Gather up the pastry sheet like a bag and twist the top like you would a garbage bag, pinching it in to seal. It should look like a miniature Santa bag. Brush with butter and bake at 350 degrees for 20 to 25 minutes, until the bag begins to brown. Place on a footed cake plate with a knife so book-club members can slice out a sliver — well, make that a hunk. They will think you are pure genius.

Special note: Don't have time to make frosting for a cake? Melt some candy bars in the microwave and pour over the top. A store-bought angel cake with melted Milky Ways will send your book-club members off to heaven or into a diabetic coma. Wait — maybe there isn't a difference between the two.

This is also not too shabby served with a couple scoops of *real* Vanilla Bean Blue Bell ice cream.

East Texas Martinis

I invented this drink at an event for author Rue McClanahan. It is now the most requested drink of the Pulpwood Queens repertoire and exactly the perfect proportion of alcohol and sweetness.

Ingredients:
Assorted tropical Popsicles
Assorted flavored vodkas (e.g., orange, cranberry, blackberry, vanilla)

Let the Pulpwood Queens and attendees select their choice of Popsicle and vodka. Unwrap their choice of Popsicle, place it in a hot pink plastic cup, stick-side up. Then measure out two shots of their favorite flavored vodka and pour over the Popsicle. Hand them their drink and tell them to "Dip and suck!" My favorite? A mango Popsicle with L'Orange Grey Goose vodka!

Pulpwood Queen DD Cupcakes

These are big cupcakes and perfect for any special occasion! I actually made these for my daughter's Jefferson cheerleader fundraiser; my daughter was the junior high school mascot. We sold them like crazy!

Ingredients:
1 box milk-chocolate cake mix
1 box vanilla cake mix
2 packages prepared cookie dough that is ready to place on a cookie sheet (1 chocolate chip cookie dough, 1 white chocolate chip with macadamia nut cookie dough)
2 extra-large containers store-bought frost-

ing (1 milk chocolate, 1 vanilla)
1 large package extra-large strawberries
1 package miniature Hershey bars
1 package Hershey white chocolate Kisses with almonds

Prepare each cake mix according to directions but substitute milk for the water and butter for the oil. Spray extra-jumbo cupcake pans with baking spray. Pour cake batter to fill individual cupcake holders in pan one-third full. Take 1 cookie dough ball, slightly flatten, and place in the center of each cupcake. Fill batter to almost full of each cupcake. Bake according to cake mix directions, then remove to cool.

When cooled, alternate milk chocolate and vanilla cupcakes with milk chocolate and vanilla icing. Place in the center of each jumbo cupcake 1 washed and dried strawberry, 1 unwrapped miniature Hershey bar, and 1 Hershey Kiss.

MISCELLANEOUS RECIPES
Chocolate Gravy and Biscuits
Here's a recipe that is worthy of our Pulpwood Queens Brunch held Girlfriend Weekend! This recipe is courtesy of Denise Hildreth, author of the "Savannah from Savannah" series and *Flies on the Butter,* but

I have found it is great at any time if you love chocolate and biscuits as much as I do.

Chocolate Gravy
Ingredients:
1 1/2 cups sugar
3 tablespoons flour
3 tablespoons cocoa
2 cups milk
1 teaspoon vanilla extract

Combine dry ingredients. Stir until completely incorporated. Whisk in milk and place over medium-high heat. Whisk continuously for about 20 minutes (or until the raw taste of the flour is cooked out). Stir in vanilla. Serve immediately over freshly baked baking powder biscuits. Should serve four.

Biscuits
I have made biscuits from scratch but nothing comes out any better for me than those frozen ones you buy in a bag at the grocery store. You slap those on a cookie sheet and bake according to directions: perfection every time. Triple your Chocolate Gravy recipe for a bag of 12 biscuits.

PULPWOOD QUEENS FOR THE NEXT GENERATION

"What are the Pulpwood Queens? Well, they are a book club, but not just *any* book club. No, they are the largest meeting book club in America. They all wear tiaras and pink shirts and leopard (except for the guys, who wear crowns and black) and read interesting and captivating books about all sorts of subjects. There are several different groups, or chapters as my mom calls them, spread out all over the country and each one is as I said different. There are some that are really big and some with only two. Some of them are really wild and crazy and some are very mild and quiet. Yet, each one is fantastically fun. My mom most of the time gets the author of the book they are reading that month to come to the meetings. So, for the first week of the month my mom hauls authors to the book club in our beat-up,

497

dented, gray minivan."

— *Helaina Amethyst Patrick, from her unpublished novel* **Memoirs of a Pulpwood Queen's Daughter**

"I saw the bright light!

I have always had a good memory, even my parents think I am gifted. I can hear a song just one time and I know the lyrics.

I can even remember when I was born. When I came out of the big pillow chamber I saw this big UFO light. I saw only one weird man with ugly glasses. I determined that he was a doctor. Then when he came home with me and my mom I realized that he was my father.

After a week or so I started to get along with my dad. I guess you say you can't judge a book by its cover, can you?

My new sister was happy to see she had a new minion. And she was a lot nicer when I was a baby. Maybe that was her evil scheme to make me like her and then when she needed something I would give it to her (I was a real pushover then)."

— *Madeleine Alexandrite Patrick, from her unpublished novel* **Memoirs of a Pulpwood Queen's Youngest Daughter**

PULPWOOD QUEEN
SELECTIONS 2000–2008

2000

March: *Divine Secrets of the Ya-Ya Sisterhood* by Rebecca Wells

April: *The Sweet Potato Queens' Book of Love* by Jill Conner Browne

May: *The House of Gentle Men* by Kathy Hepinstall

June: *World of Pies* by Karen Stolz

July: *Down from the Dog Star* by Daniel Glover

August: *Last Days of Summer* by Steve Kluger

September: *The Poisonwood Bible* by Barbara Kingsolver

October: *Life Is So Good* by George Dawson

November: *Carolina Moon* by Nora Roberts

December: *The Gates of the Alamo* by Stephen Harrigan

Bonus Books

My Louisiana Sky, Mister and Me, When Zachary Beaver Came to Town by Kimberly Willis Holt

The Mother-in-Law Diaries by Carol Dawson

The Last Madam by Christine Wiltz

Little Altars Everywhere by Rebecca Wells

The Bushwhacker by Jennifer Johnson Garrity

Mom, Are We There Yet? by Dr. Monica Anderson

Uncle Bubba's Chicken Wing Fling by Mitchel Whitington

Texas Almanac edited by Mary Ramos

So Good . . . Make You Slap Your Mama by Marlyn Monette

Robbers by Christopher Cook

The Alamo Story by J. R. Edmondson

Haunted Texas Vacations by Lisa Farwell

Killing Cynthia Ann by Charles Brashear

The Train to Estelline by Jane Roberts Woods

2001

January: *Making Waves in Zion* by Sandra King Ray

February: *God Save the Sweet Potato Queens* by Jill Conner Browne

March: *Bald in the Land of Big Hair* by Joni Rodgers

April: *Cane River* by Lalita Tademy
May: *The Florabama Ladies' Auxiliary & Sewing Circle* by Lois Battle
June: *All Over but the Shoutin'* by Rick Bragg
July: *Orange Crush* by Tim Dorsey
August: *Ava's Man* by Rick Bragg
September: *Taps* by Willie Morris
October: *The Texas Country Reporter* by Bob Phillips
November: *Blackbird* by Jennifer Lauck
December: *If Nights Could Talk* by Marsha Recknagel

Bonus Books
From My Mother's Hands by Susie Kelly Flatau
Ida Mae Tutweiler & the Traveling Tea Party by Ginnie Siena Bivona
Beauty: The New Basics by Rona Berg
Storyville by Lois Battle
Close Calls by Jan Reid
The Selling Safari by Alan C. Buhler

2002

January: *Screen Door Jesus & Other Stories* by Christopher Cook
February: *Pretty Is As Pretty Does* by Alison Clement
March: *The Absence of Nectar* by Kathy Hepinstall

April: *The Bullet Meant for Me* by Jan Reid

May: *A False Sense of Well Being* by Jeanne Braselton

June: *The Dive from Clausen's Pier* by Ann Packer

July: *Still Waters* by Jennifer Lauck

August: *Social Crimes* by Jane Stanton Hitchcock

September: *The Rich Part of Life* by Jim Kokoris

October: *The Last Girls* by Lee Smith

November: *Ruby Ann's Down Home Trailer Park Cookbook* by Ruby Ann Boxcar

December: *Christmas Past* by Bill Duncliffe

Bonus Books

Electroboy: A Memoir of Mania by Andy Behrman

The Road to Eden's Ridge by M. L. Rose

2003

January: *The Sunday Wife* by Cassandra King

February: *Prince of Lost Places* by Kathy Hepinstall

March: *Grace* by Jane Roberts Wood

April: *Catch Me If You Can* by Frank W. Abagnale

May: *A Burning in Homeland* by Richard Yancey

June: *The Canal House* by Mark Lee

July: *Somebody's Someone* by Regina Louise

August: *Fanny and Sue* by Karen Stolz

September: *Ten on Sunday: The Secret Life of Men* by Alan Eisenstock

October: *The Song Reader* by Lisa Tucker

November: *Slow Way Home* by Michael Morris

December: *Roseborough* by Jane Roberts Wood

Bonus Books

The Garden Club by Robert DeBlieux

To Kill a Mockingbird by Harper Lee

2004

January: *Fat Girls and Lawn Chairs* by Cheryl Peck

February: *Let Us Eat Cake* by Sharon Boorstin

March: *Lunch at the Piccadilly* by Clyde Edgerton

April: *The Yokota Officers Club* by Sarah Bird

May: *Some Kind of Miracle* by Iris Rainer Dart

June: *Sister North* by Jim Kokoris

July: *The Last of the Honky-Tonk Angels* by Marsha Moyer

August: *School of Dreams* by Edward Humes

September: *The Bridge* by Doug Marlette

October: *Liberating Paris* by Linda Bloodworth Thomason

November: *Bergdorf Blondes* by Plum Sykes

December: *The Grrl Genius Guide to Sex (with Other People)* by Cathryn Michon

Bonus Books

Imagine by Joyce Wilson

From Stress to Strength: Ice Cream for Breakfast by Leslie Levine

A Poetry Break by Kay Day

Alamo House, The Boyfriend School, The Mommy Club by Sarah Bird

Conversations with American Women Writers by Sarah Anne Johnson

And My Shoes Keep Walking Back to You by Kathi Kamen Goldmark

Live Like You Were Dying by Michael Morris

Defying Gravity: A Celebration of Late-Blooming Women by Prill Boyle

How to Remodel a Man by W. Bruce Cameron

2005

January: *The Same Sweet Girls* by Cassandra King

February: *The Turtle Warrior* by Mary Relindes Ellis

March: *Angry Housewives Eating Bon Bons* by Lorna Landvik

April: *Savannah from Savannah* by Denise Hildreth

May: *The Garden Angel* by Mindy Friddle

June: *Cooking for Love* by Sharon Boorstin

July: *Saints at the River* by Ron Rash

August: *Last Child in the Woods* by Richard Louv

September: *The Glass Castle* by Jeannette Walls

October: *The City of Falling Angels* by John Berendt

November: *Big Cats: Stories* by Holiday Reinhorn

December: *Last Moon Dancing* by Monique Maria Schmidt

Bonus Books

Younger by the Day by Victoria Moran

One Foot in Eden by Ron Rash

Dish by Jeannette Walls

Savannah Comes Undone by Denise Hildreth

Oh My Stars by Lorna Landvik

Texas Hold 'Em by Kinky Friedman

In the Dark of the Moon by Suzanne Hudson

Here's to You, Jackie Robinson by Joe Formichella

A Change of Heart by Philip Gulley

Bookstore Tourism by Larry Portzline

Midnight in the Garden of Good and Evil by John Berendt

The Christmas Scrapbook by Philip Gulley

2006

January: *The Life All Around Me by Ellen Foster* by Kaye Gibbons

February: *Mad Girls in Love* by Michael Lee West

March: *The Singing and Dancing Daughters of God* by Timothy Schaffert

April: *Dear Mrs. Lindbergh* by Kathleen Hughes

May: *Eating Heaven* by Jennie Shortridge

June: *The Bad Behavior of Belle Cantrell* by Loraine Despres

July: *The Tender Bar* by J. R. Moehringer

August: *One Mississippi* by Mark Childress

September: *Once Upon a Day: A Novel* by Lisa Tucker

October: *Magic Time* by Doug Marlette

November: *When Crickets Cry* by Charles Martin

December: *The Messenger of Magnolia Street* by River Jordan

Bonus Books
The Story of My Life by Farah Ahmedi
The Sound of Blue by Holly Payne

2007

January: *The Secret Memoirs of Jacqueline Kennedy Onassis* by Ruth Francisco

February: *The Flamenco Academy* by Sarah Bird

March: *Miss American Pie: A Diary of Love, Secrets, and Growing Up in the 1970s* by Margaret Sartor

April: *Rain Village* by Carolyn Turgeon

May: *The World Made Straight* by Ron Rash

June: *Heart in the Right Place* by Carolyn Jourdan

July: *Flies on the Butter* by Denise Hildreth

August: *Kabul Beauty School* by Deborah Rodriguez

September: *Family Acts* by Louise Shaffer

October: *The Camel Bookmobile* by Masha Hamilton

November: *Against Tall Odds* by Matt Roloff

December: *The Unnatural History of Cypress Parish* by Elise Blackwell

Bonus Books
Confessions of a Jane Austen Addict by Laurie Viera Rigler
The Reluctant Tuscan by Phil Doran

My First Five Husbands . . . and the Ones Who Got Away by Rue McClanahan
Blonde by Joyce Carol Oates
Eat, Pray, Love by Elizabeth Gilbert
The Worthy, Lord Vishnu's Love Handles by Will Clarke
Chemistry and Other Stories by Ron Rash
Chasing Fireflies, When Crikets Cry, Maggie by Charles Martin
Heartbreak Town by Marsha Moyer
The Bookman by Stayton Bonner
Anecdotal by J. Brooks Dann
Small Town Odds by Jason Headley
House of Plenty by Carol Dawson and Carol Johnston
Not Between Brothers by David Marion Wilkinson
The Night Journal by Elizabeth Crook
Dead Copy and *Scoop* by Kit Frazier
The Mercy of Thin Air by Ronlyn Domingue
Waltzing at the Piggly Wiggly by Robert Dalby
Body Movers by Stephanie Bond
Not Tonight, Honey: Wait 'Til I'm a Size 6 by Susan Reinhardt
Moon Pies and Movie Stars by Amy Wallen
Savannah Comes Undone and *Savannah by the Sea* by Denise Hildreth
The Stewardess Is Flying the Plane! by Ron Hogan

Healing Shine by Dr. Michael Johnson
The Lost Mother by Mary McGarry Morris
And She Was by Cindy Dyson
Queen of Broken Hearts by Cassandra King
Hot Water by Kathryn Jordan
The Art of Table Dancing by DC Sanfa

2008

January: *The Pulpwood Queens' Tiara-Wearing, Book-Sharing Guide to Life* by Kathy L. Patrick

The rest of my book-club selections are posted on my Web site at www.beautyand thebook.com.

REFERENCES

Bookstore Tourism: The Book Addict's Guide to Planning & Promoting Bookstore Road Trips for Bibliophiles & Other Bookshop Junkies by Larry Portzline, www.bookstoretourism.com.

Book Lust by Nancy Pearl.

More Book Lust by Nancy Pearl, www.booklust.com.

Leave Me Alone, I'm Reading by Maureen Corrigan.

Great Books for Every Book Lover by Thomas Craughwell.

READING GROUP GUIDE

1. I never dreamed in a kazillion years that I would ever be fired from a job. When I was fired, I handled the whole situation poorly. What would you do if this happened to you? Would you do anything different now that you have read my book?

2. I have often been judged by the way I look and by my profession. People tend to put other people in neat, little compartmentalized boxes. What if there were no box? What are your criteria for judging others? Do you judge a book by its cover?

3. I will always be a small-town Kansas girl, and I have grown proud of that fact. Even though I live in Texas — and now call it home — really Eureka, Kansas, will always be home to me no matter how many times I sing "Yellow Rose of Texas" or give the Texas "hook 'em" horns sign. What does

home mean to you, and how do you believe it shapes you?

4. *To Kill a Mockingbird* by Harper Lee is my favorite book of all time. What is your favorite book, and why? If you could make a list of your favorite books, which books would you include?

5. If you could do anything in this world without fear of ridicule or the amount of money you earn, what would you choose as an occupation, and why?

6. Obviously, since you are reading these discussion questions, you are either in a book club or love to discuss books. Do you think women in particular have embraced the book club culture? Why or why not?

7. If you are in a book club, do you think any of your fellow members would have been someone you would have naturally selected as a friend? What have you learned from other members that you would not have learned otherwise?

8. Pulpwood Queen Joyce Jackson Futch changed my outlook on life. Have you ever

met a friend who transformed you as a person? What does friendship mean to you and how are you a friend to others? Joyce also changed my outlook on death. What did you learn from reading about Joyce's story?

9. While growing up, I adored the Miss America pageant, but as a child of the seventies — and being forced into a pageant myself — I now find them demeaning. What do you think about beauty pageants? Do you think they have a future in today's society?

10. Author Doug Marlette believed that our goal in life is to reach a point where we no longer seek fame and fortune — the bright lights, big city — but that we become satisfied with a job well done. I too believe in such a dream. What do you think society as a whole can do to affirm that service above self is more important than the almighty buck and keeping up with the Joneses?

11. Teachers, librarians, and others who have mentored me have been a blessing in my life. What can we do to help recognize those individuals who believe that a child

has worth and should be educated with the highest respect regardless of race, color, religion, economic background, and where they live? What can we do as readers to help ensure that reading is seen as the highest form of entertainment where imaginations are concerned?

12. I have always felt, and now know, that creating beauty and reading are my calling in life — my life's mission. I love to make people feel better about themselves, and I love to encourage reading. There is no higher honor than to serve others. What does that statement mean to you? And why, as a society, don't we praise others who make our lives easier?

13. Name a book that changed your views on the world.

14. I believe that reading should be experienced by all the senses. I like to physically touch a book when I read it rather than read a book on-line. That is why we bring food that relates to our book selection to our Pulpwood Queens meetings, or we have the author come visit our book club and read excerpts from the book, or give a talk about the book.

Nothing brings me more pleasure than reading a book I love and then going to see the movie, musical, or play based on that book. I think that when you envelop all five senses into the reading experience, the book becomes much more pleasurable. Do you agree or disagree, and why?

15. Some people believe that reading and book clubs are just a trend in today's society. (People have also said this about leopard print, but judging from how much I have seen leopard print out there for the past thirty or forty years, I rest my case.) Do you believe that book clubs are just a trend like drinking hot Dr Pepper with lemon was back in the 1970s, or do you think reading and books are here to stay?

16. I truly believe that books saved me. I came from a family that was different. Only now, after writing this book, am I beginning to question if any family is really like what you see on *Leave It to Beaver* or *Father Knows Best.* What I do know is this: Reading helped me escape into another world. If I was scared, I could read a book. If I got too scared

reading the book, I could close the book. I was in control of my environment instead of the other way around. What has reading done for you? Now that you have answered that question, here is an even better one: What can you do to help promote reading?

ABOUT THE AUTHOR

When **Kathy L. Patrick** lost her job as a publisher's rep, she took that lemon and made margaritas: She opened Beauty and the Book, the world's only combination beauty salon/bookstore. Soon after that whirlwind success, Kathy founded the Pulpwood Queens of East Texas, a book club that became a nationwide phenomenon almost overnight. She set out to change the world, one reader at a time. Now, in candid and hilarious stories, Kathy shares her guiding principles, wildest escapades, and favorite titles. Her journey celebrates the transformative power of reading and good old-fashioned values: friendship, community, and women sharing honest advice, inspirational stories, secret recipes . . . and a prized tiara too.

The employees of Thorndike Press hope you have enjoyed this Large Print book. All our Thorndike and Wheeler Large Print titles are designed for easy reading, and all our books are made to last. Other Thorndike Press Large Print books are available at your library, through selected bookstores, or directly from us.

For information about titles, please call:
    (800) 223-1244

or visit our Web site at:
    http://gale.cengage.com/thorndike

To share your comments, please write:
    Publisher
    Thorndike Press
    295 Kennedy Memorial Drive
    Waterville, ME 04901